CAD Manager's Guidebook

CAD Manager's Guidebook

RALPH GRABOWSKI

The author's father,
Herbert Grabowski,
drafting aircraft parts
at age 18.

ONWORD PRESS
THOMSON LEARNING

Australia Canada Mexico Singapore Spain United Kingdom United States

CAD Manager's Guidebook
Ralph Grabowski

Business Unit Director:
Alar Elken

Executive Editor:
Sandy Clark

Acquisitions Editor:
James DeVoe

Development Editor:
John Fisher

Book Design:
Ralph Grabowski

Editorial Assistant:
Jasmine Hartman

Executive Marketing Manager:
Maura Theriault

Marketing Coordinator:
Karen Smith

Executive Production Manager:
Mary Ellen Black

Production Manager:
Larry Main

Production Editor:
Tom Stover

Art/Design Coordinator:
Stacy Masucci

Cover Image:
Cammi Noah

COPYRIGHT © 2002 OnWord Press.
OnWord Press is an imprint of
Thomson Learning

Printed in Canada
2 3 4 5 6 XXX 05 04 03 02 01

For more information contact
OnWord Press,
an imprint of Thomson Learning
PO Box 15015,
Albany, NY 12212-5015.

Or find us on the World Wide Web at
http://www.OnWordpress.com

All rights reserved. No part of this work covered by the copyright hereon may be reproduced or used in any form or by any means—graphic, electronic, or mechanical, including photocopying, recording, taping, Web distribution or information storage and retrieval systems—without written permission of the publisher.

For permission to use material from this text or product, contact us by
Tel   (800) 730-2214
Fax   (800) 730-2215
www.thomsonrights.com

Library of Congress
Cataloging-in-Publication Data

NOTICE TO THE READER

Publisher does not warrant or guarantee any of the products described herein or perform any independent analysis in connection with any of the product information contained herein. Publisher does not assume, and expressly disclaims, any obligation to obtain and include information other than that provided to it by the manufacturer.

The reader is expressly warned to consider and adopt all safety precautions that might be indicated by the activities herein and to avoid all potential hazards. By following the instructions contained herein, the reader willingly assumes all risks in connection with such instructions.

The Publisher makes no representation or warranties of any kind, including but not limited to, the warranties of fitness for particular purpose or merchantability, nor are any such representations implied with respect to the material set forth herein, and the publisher takes no responsibility with respect to such material. The publisher shall not be liable for any special, consequential, or exemplary damages resulting, in whole or part, from the readers' use of, or reliance upon, this material.

Table of Contents

Introduction xv

1 The Role of the CAD Manager 1

What Does the CAD Manager Need to Know? 1
    CAD Systems 1
    CAD Processes 2
    Software, Hardware, and Networking 2
    Keeping Up-to-Date 3
What's a CAD Manager Worth? 4
The CAD Department Budget 4
    Staffing Levels 4
    Ongoing Expenses 6
    Capital Expenses 6
    Representative Software Costs 6
Justifying New Systems 7
    Stage 1: Convincing Yourself 7
        The Politics of Software Releases 7
        Training 7
        Work Area and Ergonomics 7
        Division of Work 7
        Coordination 8
        Translation 8
        Visualization 8

System Administration 8
The Whim of Hardware Fashion 8
Overcoming Politics and Whims 9
Automating Repetitive Tasks 9
Hardware Selection 9
Hardware Procurement 10
Internet 10
Software 10
Stage 2: Meeting Upper Management's Objections 10
Operating System 11
Scanning and Digitizing 11
Project Life-cycle 12
Plotting 12
Stage 3: Convincing Your Staff 12
CAD Underutilization 13
Insufficient Training 14
Lack of Initiative 14
Fear of the Unknown 14
CAD Management Issues 15
Do CAD Brands Matter Anymore? 15
The Case Against CAD Management 17
The Disadvantages of CAD 18
Is CAD Dead? 19
The Travails of Archiving 20
Stay or Switch? 23
Are Vendors Helping CAD Managers? 24

2 Naming Drawings and Creating Symbols 27

A Simple Filename Convention 27
Industry Standards for File Names 28
AIA File Naming Convention 29
Discipline Codes 29
CSI Uniform Drawing System 31
Other UDS Specs 31
CalTrans Filenaming Convention 34
File Extensions 36

What are Symbols? 37
   Symbol Creation Summary 38
   Symbols in CAD 38
   How to Create a Symbol Library 38
   Sources of Symbols 44
   A/E/C CADD Symbology Libraries 45

3 Layer Names and Conventions 47

What Are Layers? 47
   Layers in CAD 47
   A CAD Package Without Layers 48
How to Name a Layer 49
   Most Common Standards Organizations 50
   Layers versus Levels 50
   Strategy #0: Do Nothing 50
   Strategy #1: The Simple Plan 50
   Strategy #2: The Plotter Plan 50
   Strategy #3: The Four-Step Plan 51
      Printing AutoCAD Layers 51
      Layer Names for In-house Drawings 52
   Strategy #4: Do What Your Client Says 52
   Strategy #5: Copy What Works, Make Minor Modifications . 52
The Struggle to Create Layer Standards 53
   Limitations of CAD Systems 53
   Conflict of the Disciplines 53
   Human Quirks 54
      Wildcarding Layer Names 55
American Institute of Architects 55
   Major Categories 56
   Groups and Subgroups 57
   Modifiers 57
CSI MasterFormat 58
CSi UniFormat 60
US Coast Guard 61
CalTrans Drawing Data Levels 63

ISO Layer Standard 64
   Mandatory Part 64
      Model versus Sheet 65
   Optional Part 66

4 Assigning Colors 69

Colors 69
   The Case for Color 69
      Light Gray Screening 70
   The Case for Monochrome 70
      Color Numbering Systems — What They Mean 72
How CAD Works with Colors 73
   Computer Color Numbering Systems 73
Where to Assign Colors 74
   Matching Colors to Pens 74
      Color for AutoCAD Users 74
      MicroStation Pen Control 75
      Plotters from Yesteryear 76
Color Standards 77
   AIA CAD Layer Guidelines 77
   USCG Civil Engineering Technology Center 77

5 Fonts and Patterns, Linetypes and Widths 79

Text Standards 80
   AIA and CSI Standards 81
   Fonts versus Styles 81
   The Development of the CAD Font 82
   Speeding up Text 82
   CalTrans Text Standard 83
   USCG Text Standard 84
   USACE Text Standard 85
      Imperial Text Heights 85
      Metric Text Heights 85
   TSTC Text Standard 86

Linetypes 87
   Hardwired versus Customized Linetypes 87
   One-Dimensional versus Two-Dimensional Linetypes 87
   Software versus Hardware Linetypes 88
   Scaling Linetypes 88
   Linetype Standards 89
   USCG Linetype Standard 89
   TSTC Linetype Standard 90
Line Widths 91
   Linewidth Standards 91
      MicroStation Lineweights 91
      Lineweights and Colors 92
   National CAD Standards Line Widths 92
   TSTC Line Width Standard 92
   USACE Line Weight Standard 93
Hatch Patterns 94
   Hatch Pattern Standards 94

6 Scale Factors and Dimensions 97

Determining the Scale Factor 98
Dimensioning With CAD 100
   Dimensions in 3D 100
   The Anatomy of a Dimension 101
   Dimensioning Standards 102

7 Standard Drawings and Templates 105

Drawing Sizes 106
   ISO (Metric) Sheet Sizes: A0 through A4 107
   ANSI Sheet Sizes: A through E 108
   JIS (Japanese) Sheet Sizes: A0 through A4 109
   DIN (German) Sheet Sizes: A0 through A4 110
Borders and Title Blocks 111
   CSI Drawing Sheet Specification 112

Placing Sheets in Order 113
   The Single Design Model 113
   Standards for Sheet Identification 114
   Nomenclature 116
Creating a Template Drawing 116
   What is a Template Drawing? 116
   Overview of Drawing Settings 117
   Preparing the Template Drawing 117
   Azimuth 118
   Template Creation Steps 118
   Graphic Scale 123
   After Templates are Created 127

8 Writing Your CAD Standards Manual 129

   Title Page 130
   Disclaimer Page 131
   Table of Contents 131
   Standard Office Library 136
   After the Document Is Complete 137

9 Working with Paper Drawings 139

   Threshholding Grayscale Scans 140
Archiving Drawings 141
   The Scanner 141
   Optical versus Software Resolution 141
   The Software 142
   Sizable Storage 143
   The Printer 144
Raster Issues and Calculations 144
   Dots Per Inch 144
   Color Depth 145
   Compression Issues 145
   Where is GIF? 145
   More Than 24 Bits 145

    The Effect of Resolution on Scan Quality 146
    JBIG2 and JPEG 2000 148
    Why You Shouldn't Use JPEG 148
Converting Drawings 149
    Original Drawings Are Not Accurate 150
    Difference between Vector and Raster 150
Two Editing Solutions 152
    The Partial Conversion Strategy 153
Case Study: The Digital Drawing 154
    Eliminating the Paper Trail 154

10 Outsourcing and Extranets 157

What is a Service Bureau? 158
    Bureau Services 159
Outsourcing Project Management 160
    Gallery of Web-based Services 161
Service Bureau Pricing 165

11 The DWG Format and Its Future 169

The DWG File Specification 171
    General Structure 171
    File Header 173
    DWG Header Variables 174
    Class Definitions 174
    Padding (R13c3 and later) 174
    Image Data (pre-R13c3) 174
    Object Data 174
    Object Map 179
    Unknown Section 179
    Second Header 179
    Image Data (R13c3 and Later) 179
    Proxy Entity Graphics 179
    EOF 181

The Future of File Formats 182
    Objects in CAD 182
    From Files to Central Databases 184
    Working with Objects 185
    The Downside to Objects 186
    Server-based Systems 187
    ASP-based Systems 189
    The Future of CAD Is 190
    The Aging of Software 192
    Conclusion 192

12 A Recent History of CAD 195

The Beginnings of CAD 196
    1995 197
    1996 202
    1997 208
    1998 216
    1999 224
    2000 230
    2001 236

Appendices

A Resources for CAD Managers ... 241

Starting Points ... 241
CAD Vendors ... 243
   CAD for Windows ... 243
   CAD for Macintosh ... 248
   CAD for Linux ... 249
Standards Bodies ... 251

B Color-Pen Table ... 253

   Resource ... 254

C The CSI Layer Standards ... 263

MasterFormat ... 263
   Division 0 : Introductory Information, Bidding, and Contracting Requirements ... 263
   Division 1: General Requirements ... 264
   Division 2: Site Construction ... 264
   Division 3: Concrete ... 264
   Division 4: Masonry ... 265
   Division 5: Metals ... 265
   Division 6: Wood and Plastics ... 265
   Division 7: Thermal and Moisture Protection ... 265
   Division 8: Doors and Windows ... 266
   Division 9: Finishes ... 266
   Division 10: Specialties ... 266
   Division 11: Equipment ... 267
   Division 12: Furnishings ... 268
   Division 13: Special Construction ... 268
   Division 14: Conveying Systems ... 269
   Division 15: Mechanical ... 269
   Division 16: Electrical ... 269

UniFormat 270
   10: Project Description 270
   20: Proposal, Bidding, and Contracting 270
   30: Cost Summary 270
   A: Substructure 270
   B: Shell 270
   C: Interiors 271
   D: Services 271
   E: Equipment and Furnishings 272
   F: Special Construction and Demolition 272
   G: Building Sitework 273
   Z: General 274

US Coast Guard Master Listing 275
   Division 01: Field Engineering 275
   Division 02: Sitework 275
   Division 03: Concrete 275
   Division 04: Masonry 276
   Division 05: Metals 276
   Division 06: Wood and Plastics 276
   Division 07: Thermal and Moisture Protection 276
   Division 08: Doors and Windows 277
   Division 09: Finishes 277
   Division 10: Specialties 277
   Division 11: Equipment 277
   Division 12: Furnishings 277
   Division 13: Special Construction 277
   Division 14: Conveying Systems 278
   Division 15: Mechanical 278
   Division 16: Electrical 279
   Division 20: Reference 279

Index 281

Introduction

The client was on the phone, breathless:

"We've installed our first CAD station. Now we want to begin our first drawing. What colors do we use? What do we name the layers? How do we name drawing files and which subdirectories do we put them in?"

"Anything you want," I answered, at first.

. . .

That's the problem with CAD (computer-aided design) systems: most are so flexible that you create a drawing anyway you want. Draw with any of 16 million colors; give layers and files any name up to 255 characters long; and store drawings in any folder on the computer's hard drive, a co-worker's hard drive (via the network), burn it onto a CD, or store it somewhere on the Internet.

Fortunately, the client was smart enough to know that his firm needed to define layer, color, and file-naming conventions *before starting the first drawing.*

My second, more realistic answer was to tell him the choices he had in selecting a system for organizing drawings. In this book, you learn about conventions that help you get started on your first drawings. Other chapters introduce you to concepts you need to consider to manage a CAD system.

> **ABOUT THE SIDEBARS**
>
> In addition to the hardcore data on the pages of this book, you'll notice a lot of extra information in the "sidebars" of this book. To make the book more fun and useful, look for tips and CAD paraphernalia filling the sidebars.

THE AUDIENCE FOR THIS BOOK

This book is meant for the "CAD manager," the person in charge of CAD in your office. You might have a number of names — Manager of CAD Systems, Computer Manager, Information Systems, or just "the guy who knows the most about CAD."

The purpose of this book is to help you dictate some order among the disorder that CAD can create. There are a lot of ideas for naming layers and drawings, placing sheets in the correct order, understanding the past, and thinking about challenges that may face CAD management in the future.

The rules for running CAD in an office are much the same for any brand name; just the terminology and some features are different. For this reason, I assume you have a CAD system — *any* CAD system. This book discusses CAD in a generic fashion, yet makes specific references to some CAD packages, such as AutoCAD, MicroStation, Cadkey, Revit, ArchiCAD, and others.

While Autodesk's AutoCAD has become a dominant package among CAD software, increasing compatibility with AutoCAD makes competitive CAD packages worth considering. In today's market, you cannot afford to be loyal to any one vendor.

THE ORGANIZATION OF THIS BOOK

The CAD Manager's Guidebook is divided into three sections: an introduction to CAD management, developing a CAD standard, and other CAD topics.

Introduction to CAD Management

Chapter 1: The Role of the CAD Manager. While the CAD manager's role appears to be one of making sure your firm's CAD system is running smoothly, much of your time is spent justifying additional purchases of hardware and software, solving users' problems, and dealing with office politics.

You find yourself facing a host of problems, and this chapter describes solutions to problems such as unrealistic expectations by management and the need for on-going training. This chapter also discusses in frank terms CAD management issues.

Creating a CAD Standard

Chapters 2 through 8 describe numerous CAD standards, and how to create a standard for your office.

Chapter 2: Naming Drawings and Creating Symbols. Extensive use of symbols is the key to increasing CAD productivity, however, you must create a system of naming and organizing libraries of symbols. This chapter describes systems of naming symbols and drawing files.

Chapter 3: Layer Names and Conventions. A number of semi-official organizations have created standards for naming layers. This chapter describes several standards, as well as gives step-by-step instructions for creating your own in-house standard.

Chapter 4: Assigning Colors. It used to be that the purpose of an entity's color was to select the correct pen during plotting. With the demise of the pen plotter and its replacement of the inkjet and laser plotter, colors are no longer strictly for controlling plots. This chapter describes how CAD works with colors, as well as several conventions for the use of colors in CAD drawings.

Chapter 5: Fonts and Patterns, Linetypes and Widths. With the advent of CAD systems that can read TrueType font files, drawings have access to thousands of fonts. This chapter warns against the "ransom note" look and shows how to create a standard with just one font and four sizes of text. It also describes standards for fonts, hatch patterns, linetypes, and line widths.

Chapter 6: Scale Factors and Dimensions. This chapter shows how to scale drawings, and the calculations required. Standards for dimensions are also described.

Chapter 7: Standard Drawings and Templates. Most CAD packages include international standard drawings, complete with drawing border and title block. These are illustrated in this chapter. You learn about the order in which to place sheets to create a drawing set. With all the elements in place, you can create a prototype drawing that ensures that all your firm's drawings are made to the same standard.

Chapter 8: Writing Your CAD Standards Manual. Finally, you should document your firm's CAD standards. This chapter shows how to organize the standard in written form, either in a three-ring binder or on your firm's intranet Web site.

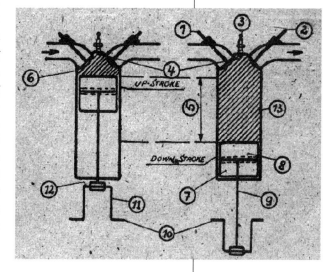

Maximizing CAD Efficiency

After setting the CAD standards for your office, there are ways to make CAD work more efficiently.

Chapter 9: Working with Paper Drawings. This chapter describes how to get existing paper drawings into your CAD system. It discusses whether you need to deal with paper at all by going "all digital."

Chapter 10: Outsourcing and Extranets. Consider service bureaus, whether local or located elsewhere in the world, as an "office overload" center. This chapter describes the services provided by service bureaus and the charges for typical services, and it gives advice on dealing with bureaus, including those located on Web sites.

Chapter 11: The DWG Format and Its Future. Getting more technical, this chapter provides a condensed listing of the contents of a DWG drawing file created by AutoCAD. You read about what the future of CAD files might be, and how that might affect your office.

Chapter 12: A Recent History of CAD. To better understand the changes that are bound to occur in the future, it helps to know what when on in the past. This chapter describes the six years of CAD history, beginning in 1995.

Appendices

Additional information is provided in the appendices:

Appendix A: Resources for CAD Managers. Web sites for additional CAD resources, CAD vendors, and standards bodies.

Appendix B: Color-Pen Table. Matching 255 drawing color to pens for use with AutoCAD and MicroStation.

Appendix C: The CSI Layer Standards. A more detailed look at the CAD layer guidelines supported by Construction Specifications Institute and adapted by Uniformat and the United States Coast Guard.

TERMINOLOGY USED IN THIS BOOK

Since this book is not specific to one CAD package, the problem of terminology arises. While CAD packages agree on the use and meaning of the term "line," they have different terms for just about everything else. Is it a "drawing" or a "design file"? Is it a "symbol" or a "block" or a "component" or a "group" or a "part" or a "cell"?

This book tries to use generic terminology (such as "symbol") but does tend to lean toward AutoCAD terminology such as "layer." Since AutoCAD is currently the most pervasive CAD package, more readers will be familiar with AutoCAD terms than of any other. Occasionally reference is made to the jargon of a specific CAD program.

ON THE COMPANION CD

The CD included with this book contains references useful for creating CAD standards. Each reference is in its own folder:

AEC CAD Standard: The *A/E/C CAD Standard* document (in PDF format) from the Tri-Service CAD/GIS Technology Center and published by United States Army Engineer Research and Development Center.

Details Library: The *CADD Details Library, Report 1: Architectural Details* document (in PDF format) from the Tri-Service CAD/GIS Technology Center and published by United States Army Engineer Research and Development Center.

CE CAD Manual: The *Applications and Standards Manual* document (in PDF format) for CE-CADD published by the United States Coast Guard Civil Engineering Technology Center.

Developer's Guide: The *A Developer's Guide for Producing and Publishing Engineering Documents* (in PDF format) published by the United States Army Corps of Engineers.

OpenDWG Doc: The *AutoCAD R13/R14/R2000 DWG File Specification* (in RTF format) published by the OpenDWG Alliance.

About the Author

Ralph Grabowski received his BASc degree in engineering from the University of British Columbia in 1990. He publishes *upFront.eZine*, a weekly e-newsletter that covers the business of CAD.

Mr. Grabowski was the first Technical Editor at *CADalyst*. He also worked as a Contributing Editor to *Cadence* magazine, the Launch Editor for *Technical Design Solutions* magazine, and is currently Editor of *AutoCAD User* magazine.

Mr. Grabowski has written over fifty books on computer-aided design and graphics including *The Illustrated AutoCAD 2002 Quick Reference* (Autodesk Press), *Learn Visio 2000 for the Advanced User* (WordWare), and he is the coauthor of *MicroStation for AutoCAD Users* (OnWord Press).

You can reach Mr. Grabowski via e-mail at ralphg@xyzpress.com and visit his Web site at www.upfrontezine.com.

Acknowledgments

First and foremost, I must thank Don Beaton for his technical editing of this book. Don has the benefit of being the CAD manager for a large engineering firm in California. He not only corrected and questioned my text, he also contributed significant portions.

Thank you also to Pamela Lamb, whose eagle eye caught more copy errors than I care to admit. Thanks, too, to John Fisher at Delmar Thomson Learning (a.k.a. OnWord Press), who guided this book through its many stages.

Then there are the many people who allowed me to use some of their material: Karl Davies, Herbert Grabowski, Volker Mueller, Evan Yares of OpenDWG Alliance, Rob Berry of IMSI, Martyn Day of *CADDesk AEC*, and the Construction Specifications Institute.

Thank you to my wife, Heather, and my children, Stefan, Heidi, and Katrina, for making time for me to finish this book. *Soli Deo gloria!*

Ralph Grabowski
Abbotsford, British Columbia
April 30, 2001

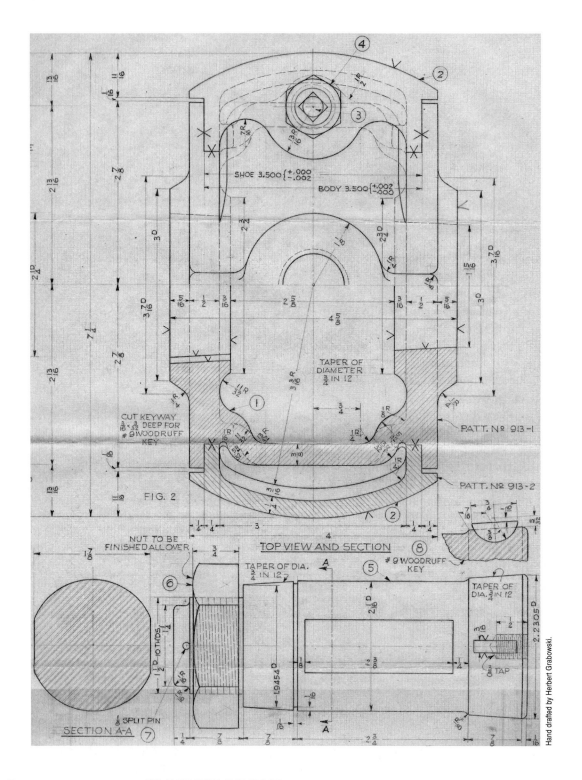

The Role of the CAD Manager

1

As the manager in charge of computer-aided design (CAD), your role is to ensure the maximum efficiency of the CAD system — even though you will be plagued daily by nitpicky problems, such as dry inkjet cartridges.

The successful CAD manager is able to ease CAD into the office by carefully planning the process and keeping everyone informed. While the primary problem you will face is finding money for upgrades to the system, above all remember that the computer is simply a more efficient tool for getting the firm's work done.

WHAT DOES THE CAD MANAGER NEED TO KNOW?

The easiest definition of a CAD manager is "anyone who knows more about CAD than you do."

Seriously, the CAD manager is a combination of an excellent CAD user and a network manager. Some of the things with which you should be familiar are:

CAD Systems

You must be familiar with all aspects of the CAD system your firm uses:
- Drawing and editing commands and techniques.
- Plotting and archiving drawing files.
- Customizing and programming.

CHAPTER SUMMARY

This chapter describes the job of CAD manager, and answers these questions:
- What does the CAD manager need to know?
- What's a CAD manager worth?
- How much for the CAD department's budget?
- Can CAD be underutilized?
- Are vendors helping CAD managers?
- Is there a case against CAD management?

- Translation to and from other CAD systems and data formats.
- Understanding software related to CAD, such as databases and VBA (Visual Basic for Applications) programming.

In addition, it's helpful to have some familiarity with other CAD systems. Some firms use more than one CAD package in-house, and you will definitely have at least one client using a different system. In any case, getting to know a second CAD system is like learning a second language: it broadens your horizons.

CAD Processes

CAD is more than drawing, editing, and plotting drawings. There is a whole range of processes that accompany the production of drawing. These are the tasks most closely identified with the profession of CAD management, such as:

- Implementing a document management system that controls access to drawing files.
- Preparing a revision management system so that everyone knows which version of the drawing is the most recent.
- Making CAD standards and standardized drawings (symbols, title blocks, borders, etc.) available from a central location to all drafting stations.
- Having the ability to transmit drawings to clients electronically.
- Setting up an on-line project management system, perhaps with group discussion capability.
- Justify to upper management the implementation of new technology.
- Hiring (and, sometimes, firing) staff.

This book deals with many of these items.

Software, Hardware, and Networking

CAD runs on computers, and this involves a set of skills unknown to the manual CAD drafter. Many of the following tasks are usually assigned to the IT (information technology) department, the network manager, or in a small firm, to the CAD manager:

- Installing software, and adding hardware components.
- Connecting and configuring the network.
- Tweaking the operating system, and downloading the latest drivers and software patches.
- Understanding how peripherals work.

- Setting up backup, anti-virus, and anti-cracker systems.
- Knowing how to use the Internet, including making connections, e-mail, FTP (file transfer protocol), hyperlinking, and Web searching.

Keeping Up-to-Date

CAD continues to change every few months, and it is necessary for the CAD manager to keep up-to-date via:

- Employing training, whether self-taught or in a classroom via CD-ROM, videotapes, or books.
- Reading industry magazines, including business magazines, discipline-specific publications, general computer magazines, and publications specific to CAD.
- Monitoring informational Web sites, e-newsletters, and CAD vendor newsgroups.
- Maintaining CAD standards for your office.
- Keeping track of the latest trends in the CAD world, and having an understanding if they are genuinely useful or just hype being used to sell a new product.

Fortunately, the Internet makes it much easier to stay up-to-date with the latest technology. For example, I spend roughly an hour a day reading news and opinion at Web sites, including these:

my.yahoo.com is customized for the world news, tech news, weather, sports, and stocks that I am interested in.
www.cnet.com provides news about the computer industry, and is updated throughout the day.
www.theglobeandmail.com has news of Canada, where I'm located.
asia.cnn.com provides news about Asia and the world.
quote.yahoo.com/m3?u tracks the exchange rates of currencies.
www.theregister.co.uk gives the British slant (and "humour") to the daily computer news.
betanews.efront.com has latest news about unreleased software.
www.ugeek.com/pdageek covers news and reviews of personal digital assistants, such as the Palm.

In addition, I subscribe to many e-newsletters and magazines, which are delivered by e-mail every day, weekly, biweekly, or monthly.

A list of topics and replies in a typical Web-based discussion forum. This one is hosted at the www.TurboCAD.com Web site.

WHAT'S A CAD MANAGER WORTH?

Cadence magazine did CAD managers a service with its pay survey,[*] which answered the question, "How much should you be paid?"

The survey showed that the range of salaries is $38,336 to $52,712, with pay tied to the person's abilities and responsibilities. (Amounts in stated in United States funds.) Over half of the survey respondents knew how to program. The pay tended to be lower in smaller cities.

The typical CAD manager is responsible for between 2.7 and 16.1 CAD positions. Only 55 percent of respondents to the survey had purchasing power. There seemed to be a relationship between personnel and budget control: if a CAD manager was responsible for just the computers, but not staff, another manager usually controlled the budget.

The number one CAD package was AutoCAD, outdistancing other software by a margin of 30-to-1. The AutoCAD bias is probably due to the tendency of AutoCAD users to read the magazine. Still, with the popularity of AutoCAD, it won't hurt your career to know that particular program, even if your current employer doesn't use it.

For educational background, a CAD manager typically has five years experience with engineering or architecture and has a technical diploma or university degree. The CAD manager sometimes gains his position by default, since he was the resource person — someone who knew the answers to the questions from others in the department.

[*] *Cadence*, November 2000: "Manager's Viewpoint: CAD Manager Demographics."

THE CAD DEPARTMENT BUDGET

When working on the annual budget for your CAD department, take into consideration the following factors:

Staffing Levels

If you plan to hire additional drafters, remember to include the cost of new workstations, software licenses, and training.

If you plan to let go of staff, you might be able to make use of the freed-up salaries for improving the CAD department's hardware and software infrastructure. A number of years ago, the head of an industrial drafting department told me he was looking forward to the retirement of an employee, because his wage would be going toward networking the department.

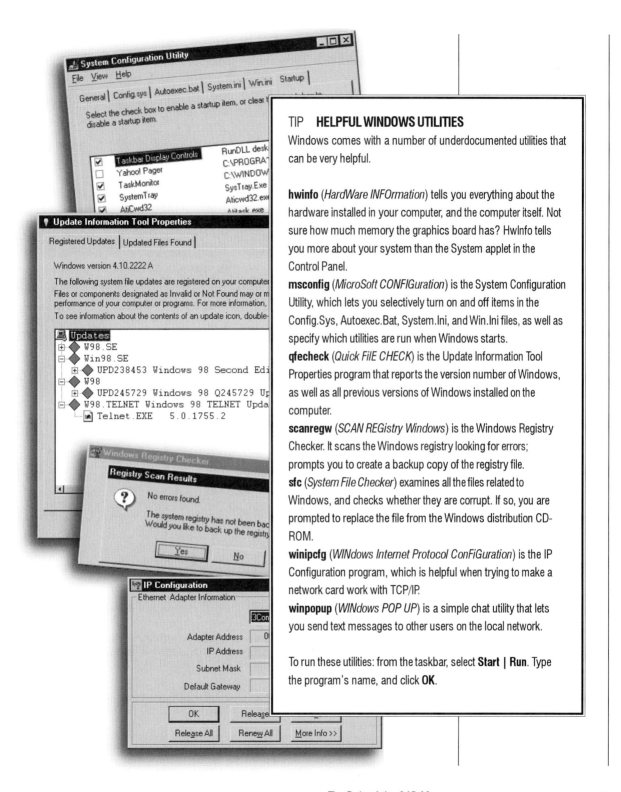

TIP **HELPFUL WINDOWS UTILITIES**

Windows comes with a number of underdocumented utilities that can be very helpful.

hwinfo (*HardWare INFOrmation*) tells you everything about the hardware installed in your computer, and the computer itself. Not sure how much memory the graphics board has? HwInfo tells you more about your system than the System applet in the Control Panel.

msconfig (*MicroSoft CONFIGuration*) is the System Configuration Utility, which lets you selectively turn on and off items in the Config.Sys, Autoexec.Bat, System.Ini, and Win.Ini files, as well as specify which utilities are run when Windows starts.

qfecheck (*Quick FilE CHECK*) is the Update Information Tool Properties program that reports the version number of Windows, as well as all previous versions of Windows installed on the computer.

scanregw (*SCAN REGistry Windows*) is the Windows Registry Checker. It scans the Windows registry looking for errors; prompts you to create a backup copy of the registry file.

sfc (*System File Checker*) examines all the files related to Windows, and checks whether they are corrupt. If so, you are prompted to replace the file from the Windows distribution CD-ROM.

winipcfg (*WINdows Internet Protocol ConFiGuration*) is the IP Configuration program, which is helpful when trying to make a network card work with TCP/IP.

winpopup (*WINdows POP UP*) is a simple chat utility that lets you send text messages to other users on the local network.

To run these utilities: from the taskbar, select **Start | Run**. Type the program's name, and click **OK**.

The Role of the CAD Manager

> **TIP GETTING HELP WITH WINDOWS**
>
> Microsoft is not fond of helping customers, unless we're willing to pay. It can be difficult to find answers with their free on-line "knowledge-base" at www.microsoft.com/support.
>
> The solution, I have found, is to type the error message or my question at the www.google.com search engine. I often find the answer within a couple of clicks.
>
> After finding the solution, print it out, and save in a three-ring binder.

Ongoing Expenses

These tend to be predictable, and include magazine and software subscriptions, support contracts, office supplies (such as inkjet cartridges), replacement of low-cost hardware (such as a worn-out keyboard), and ongoing training.

Capital Expenses

Large spending on hardware and software are called "capital expenditures" because of they are taxed by the depreciation method: their cost is written off over a multi-year period, according to the schedule set out by your government.

In Canada, for example, it takes two years to write off software from the corporate income tax. Ask your firm's accounting department about the best way and time to make large purchases. The best way might be leasing; the best time might be at the end of your firm's fiscal year.

Representative Software Costs

Here are some examples of the cost of software licenses taken from the www.autodesk.com/estore Web site, Ketiv Technologies (an authorized AutoCAD dealer), and other sources. Use these as examples only; prices vary due to many factors, such as the purchase source, and changes by the vendor (prices are in United States funds and are rounded to the nearest $10).

	AutoCAD	**AutoCAD LT**	**ArchDesktop**	**Inventor**
Onetime license[1]	$2,900[2]	$660	$3,400[3]	$5,000
Upgrade previous version	$400	$200	$350	$1,300
Upgrade older version	$700	$200	$600	N/A
Incremental release[4]	$150 ea	N/A	$150 ea	N/A
Annual subscription[5]	$150/yr	N/A	$500/yr	$900/yr
Corporate subscription	From $5,000 to $45,000 per site			

<u>Notes</u>:

[1] Lower prices available for larger purchases; e.g., purchase five units and save 10%. Governments, educational institutes, and large corporations receive lower prices.
[2] Dealer price; Autodesk Web price is $3,295.
[3] Includes AutoCAD 2000i.
[4] Price can vary; called "Extensions" by Autodesk; three to four releases a year.
[5] The annual subscription rate varies depending on when it is purchased; called "VIP."
N/A Not available for this product.

JUSTIFYING NEW SYSTEMS

Unlike drafting boards, CAD systems are not static. Every few months, a CAD vendor makes another breakthrough; every month, new hardware appears on the pages of the computer press. One of your jobs as CAD manager is to justify new purchases. There are three stages to this:

Stage 1: Convincing yourself.
Stage 2: Convincing upper management.
Stage 3: Convincing your staff.

Stage 1: Convincing Yourself

In stage 1, you need to convince yourself that the new gee-whiz feature is worth the cost of upgrading. Even if it is free, there is still the cost of time — researching, installation, learning the new system, and fixing the screwups that may occur. Let me digress, and write for a few moments on the politics of software releases and the whims of hardware fashions.

The Politics of Software Releases

As I write this book, CAD vendors are trying to make the switch from so-called "big R" releases to a stream of incremental releases. A big-R is a major software release that comes out every year or two, such as AutoCAD 2000, MicroStation/J, and Cadkey 19. An incremental release comes out every few months, and often consists of bug fixes plus a couple of new features.

The primary reason CAD vendors want to change from big-R to incremental is cash flow. In the past, CAD vendors would experience a huge influx of cash every eighteen months or so, as users paid to upgrade for the new release; in the months following the major release, the cash flow tapers off. When the CAD company is publicly traded (as most are), the stock market takes a dim view of a company whose revenues appear to be declining. For the benefit of the company's shareholders then, it's a bad idea to have a big-R every eighteen months.

Incremental releases generally come out every three months. (It's no coincidence that the release schedule matches the corporate financial quarter.) Users pay a smaller amount, but pay it more often, resulting in a smoother cashflow for the CAD vendor, along with greater approval from Wall Street — hopefully. To further

---

Here are aspects to managing a CAD department considered important by CAD manager Don Beaton.

TRAINING
Why train?
Who to train:
   Managers
   Engineers
   Technicians
   Drafters
Mentoring:
   Newcomer next to expert
   Open area
Methods
   Video
   Books
   Computer-based
   Instructor-lead
   In-house
   Out of house

WORK AREA & ERGONOMICS
Mouse, digitizer
Keyboard
Monitor
Chair
Desk

DIVISION OF WORK
Drafters
Technicians
Engineers
Clerical

(continued next page)

(managing CAD)

COORDINATION
Client
Consultants
Inhouse
Disciplines
Geographic

TRANSLATION
DGN
DWG
DXF
PDF
Raster to vector
Raster to raster

VISUALIZATION
Rendered image
Animation
Interactive fly-through
Panorama

SYSTEM ADMIN
Backup & restore
Tape rotation:
   On-site/off-site
Archiving
Inventory:
   Hardware
   Software
Licensing issues:
   UCITA
   Per server/per user
   Home use
   Software piracy
Upgrades:
   When to upgrade
   Maintenance contract
   Pay for each upgrade

---

smooth the cash flow, the CAD vendor offers subscriptions, for which you pay an annual fee to "belong to the club," get all incremental releases free, and presumably pay less than buying the increments individually.

There is, however, a problem with incremental releases. Not all users are going to be interested in switching payment plans if they are used to budgeting for the big-R every two years. Not all users care to update their CAD system every three months. This leaves the CAD user base at different stages, and a big-R release is required to synchronize users — as SolidWorks and Bentley Systems found out.

On the positive side, you may prefer a small, incremental upgrade over one large revision that negatively affects your workflow. There may also be tax advantages in paying for a subscription over paying for a software license.

The Whim of Hardware Fashion

The computer revolution is over. It ended the day Apple released its translucently colored iMac computers. Instead of buying a computer for its CPU speed, consumers now purchase a computer for its looks. That day had to come; never mind Moore's Law, which states that Intel will be designing multi-pentahertz CPUs at some point in the not-so-distant future. Because computers are finally fast enough.

In the early days of the personal computer, you bought the faster computer because you had to. The first IBM PC (using a 4.77 MHz 8-bit CPU) regenerated AutoCAD's famous Nozzle drawing in 4.5 minutes; that was a long time to wait to change the zoom. Adding the optional math chip (FPU) reduced the time to a half-minute. By the time the 32-bit Pentium CPU came along ten years later, even the slowest 60 MHz Pentium could redraw the Nozzle as fast as you could press the spacebar (to repeat the Redraw command). Similarly, doing hidden-line removal on a 16MHz 16-bit 80386SX CPU computer would take all weekend. Today, we work in real-time hidden-line mode.

For the most part, computers are fast enough; no longer is there a need to upgrade every two or three years. My own 400 MHz Pentium II computer is sufficient after three years. There are, however, some exceptions. Perhaps the worst example is graphics boards for 3D action games; new features make the games more realistic. When a game cannot access a special effect (such as fog or reflection) on the graphics board, the effect fails to appear in the game; in some cases, the game fails to run, and a new graphics board has to be purchased. This particular problem, fortunately, does not apply to CAD.

But the CAD manager can sometimes be affected by how technology advances and ebbs. In the mid 1990s, Iomega's 100 MB Zip disk cartridges became a standard means of delivering and storing large files. Other Iomega products fared less well. The miniature 40 MB Clik drive looked cute, but suffered from undercapacity. The high-end Jaz cartridge stored 2GB of data at high speed, but was expensive compared to other storage technology. In addition, Iomega was dogged by user complaints of product unreliability; reliability is crucial for data backup and storage!

By the end of the 1990s, magnetic disk products became passe for data delivery when they were replaced by high-speed Internet connections. A good ADSL connection transmits data at about 4 MB per minute — hours faster than overnight courier delivery! For data storage, recordable CD-R discs took over in popularity when the price of discs dropped below $1 each. In addition, CD-R discs were inherently more reliable and longer lasting than magnetic media.

Overcoming Politics and Whims

So, how do you sift between the wheat and the chafe of software and hardware vendor's claims for their products? There are several sources:

- **Check the on-line discussion forums**, which many vendors' Web sites offer. The forum is a place where users voice the complaints, and urge the vendor to improve the product. It's also a good place to see how responsive the vendor is to user comments.

- **Check the vendor's Web site** for additional information; some post the user manual for the product, which provides you with hard information not found in marketing material. Many vendors post updated drivers or other helpful software at their Web site; if there is no "downloads" area, this may be a clue to avoid the vendor.

- **Read product reviews** in print and on-line publications. Reviews tend to be positive, so pay careful attention to neutral and negative comments. For example, I reviewed many pen plotters back in the 1980s. Almost all of them would have worked just fine in

---

Identifying storage hogs:
   Finding duplicate files
Security
Viruses
File deletion
File permissions:
   Users
   Groups
   Access rights

AUTOMATING REPETITIVE TASKS

custom programming
AutoCAD:
   AutoLISP
   Menus & toolbars
   Macros
   ObjectARX
   DIESEL
   PGP file
MicroStation:
   User commands
   ASCII command file
   MicroStation Basic
   MDL
   Java MDL
Operating system:
   Batch files
   Perl
   GUI task automation

HARDWARE SELECTION

Hardware requirements
Desktop:
   Minimum
   Optimal
Mobile devices
Palm devices
Network servers

(continued next page...)

(managing CAD)

HARDWARE PROCUREMENT

Choices:
    Buy
    Rent
    Lease

Brand:
    Homebrew
    Local manufacturer
    National brand name

Selection criteria:
    Low bid
    Components
    Support

Maintenance:
    Repair response time
    On-site repair
    Factory repair
    Do-it-yourself
    Vendor technician

INTERNET

E-mail
FTP
Project Web site:
    Inhouse
    Out of house
Resources
Security
Telecommuting

SOFTWARE

File compression
File management
File viewing
File format translation
Vectorizing raster
Task automation

---

any office. But one had a fatal design flaw: when the plotter ran out of paper, the pens wrote on the grit roller wheel (which moved the paper back and forth), destroying the pen tips. Not only was the roller wheel in the wrong location, but the plotter was incapable of sensing when it had run out of paper.

- **Check out the free trial version**. If the new product is software, most vendors have a free trial version that will give you hands-on experience to help you decide whether the product is worthwhile or not. The trial versions are either fully functional but run for just 15 to 30 days, while other trial versions have no time limit but disable the save and print functions. For example, I downloaded the demo version of a raster program that I use a lot, but found none of the new features were of interest to me; I'm sticking to the old version for now, and will see what's in the following version.

To emphasize this last point, many companies have a policy of skipping upgrades, or by the more official sounding "delayed implementation." For example, a significant number of AutoCAD Release 12 users skipped Release 13, then upgraded to Release 14. This allowed them to spread the upgrade cost over 3½ years, as well as reduce the hidden cost associated with upgrading. CAD vendors are not happy with this practice, but it is your right to not purchase.

Stage 2:
Meeting Upper Management's Objections

After you have convinced yourself of the usefulness of a new product, you face stage 2. In convincing upper management, keep in mind their primary objective: cost justification. They will want you to answer the question: "Does this enhancement make financial sense?" Here are some objections you should be prepared to answer:

"The current technology is good enough." Recognize that upper management is comfortable with the current technology; moving to a different technology means moving out of their comfort zone. Acknowledge their feelings by reminding them that it took them months or years to become comfortable with the current technology; adapting to the new technology will be no different. Describe

the change in evolutionary terms (incremental change), rather than revolutionary (huge change).

"No one else is using the new technology." This argument is a good one; you should not be living on the bleeding edge of technology. Avoid introducing new technology until the need is proven, such as your clients or competitors are using it. There was no point being the first to own a fax machine; there is no point upgrading software if none of your clients have it.

"We don't understand it." This objection will be cloaked in questions that show management's lack of understanding. For example, I once gave a presentation to an engineering consulting firm on the pros and cons of adding a second CAD system. As I was explaining that MicroStation has difficulty translating AutoCAD polylines, one of the firm's principles asked, "What is a polyline?" (By the way, the firm wanted to add MicroStation since their largest clients was using it.) Suggest a pilot project (perhaps using a free copy of the downloadable demo) to help management see the benefits in concrete terms.

"What are the benefits?" Management understands that there is a cost associated with change. You need to explicitly define the benefits. Benefits might include: lower ongoing costs; standardization with clients or industry; faster output; greater production; less repetition; reduction in the amount paid to outside contractors; increased capabilities, such as software-based stress analysis or heat-loss calculations. Set up an ROI table which shows the return on investment over a multiyear period. Identify the costs as best as you can, including the cost of:

· The new technology.
· Additional training.
· Production lost while training.
· Outside help, including help from other departments.
· Upgrades of associated products. For example, if you switch to MicroStation, you might need to change (or purchase new) add-on software that works only with AutoCAD.
· Temporarily reduced productivity while the new technology is being learned.

Task scheduling
Programming
Plotting:
   Vector
   Raster
   Mixed
   Batch
Security
Antivirus
Backup
CD-R creation
Personal organizer
E-mail
FTP
Web server
Scheduling
Screen capture
Text editor
GPS

OPERATING SYSTEM
Mac
Windows 9x
Windows Pro 200x
Linux
Unix
PalmOS

SCANNING & DIGITIZING
Digitizing old plans
Scanned image
Raster manipulation tools
Raster file formats

(continued next page)

(managing CAD)

PROJECT LIFE-CYCLE
Project standards
Organization
Production
QA
Delivery
Archiving

PLOTTING
Color versus B/W
Paper sizes:
   Imperial
   Metric
Media types:
   Mylar
   Vellum
   Bond
   Thermal
   Plot to file
Plotter technology
Reproduction:
   Microfilm
   Plotting to file
   PDF
   HPGL
   PostScript
Archival quality:
   Media deterioration
   Paper
   Digital
   Cost comparison
   Speed comparison

Stage 3:
Convincing Your Staff

After upper management has approved the technology, the final step is to convince your staff to use the new technology. Perhaps the most important step in making the transition to new technology is to keep everyone fully informed. The change should not come as a surprise to your staff, as the boxes land on their desk. Eliminate the agitation created by rumors and the uncertainty of employees' future employment possibilities. Managers who think that the Cloak of Secrecy is the best wardrobe selection find later they are wearing the Emperor's New Clothes as the office erupts in rebellion against the new technology.

Here are some practical steps to planning the transition to new technology based on the experiences of several firms:

- Realize that introducing new technology is a transition process. Your firm will not be able to afford the new technology on every desk right away.
- Take advantage of the fact that only some employees will initially have access to the new technology. Find out which employees are excited about getting their hands on it and train them first. In most cases, their excitement will be transferred to fellow workers, lessening the resistance to computerization. In addition, you will find the keen employees become natural mentors to the less-keen peers.
- Teach the employee that they are "information managers" who are aided by the new technology. Determine how data flows throughout the office and interfaces with other departments and with clients.

In some extreme cases, you may need to let go of staff that is unwilling to accept the changes. I worked for an engineering firm in pre-CAD times where engineers drew the plans in pencil; the drafters merely traced over the plan with ink on Mylar to create the final drawing — in fact, the drafters were sometimes referred to as "tracers." When CAD was introduced to the firm, the engineers didn't need tracers anymore, since the engineers produced the final drawings themselves. While some drafters were trained as technicians, others were let go.

CAD UNDERUTILIZATION

Even if you computerize your drafting, don't expect drafters to use their CAD workstation 100 percent of the time. A rule of thumb is that drafters and designers use CAD just 50 percent of the day; the other 50 percent of the time is spent with word processing (specs, correspondence, and contracts), spreadsheets (area calculations, budgets, and staffing), and databases (schedules, production, and construction).

According to a study by the American Institute of Architects, CAD accounts for about half of the chargeable design process. Their breakdown of services looks like this:

Phase	Designated Service	Percentage of Fee
1	Predesign	0%
2	Site analysis	0%
3	Schematic design	15%
4	Design development	20%
5	Construction development	40%
6	Bidding and negotiations	5%
7	Construction contract administration	20%
8	Post-construction	0%
9	Supplemental	0%

Cadence magazine columnist Robert Green speaks of the *technology gap* between what technology can do for you versus what you are doing with the technology. He gives the example of buying software capable of 3D design, yet using it to perform 2D drafting.

Your staff is underutilizing CAD when they don't take advantage of the software's capability. This is not unique to CAD software; the rule-of-thumb is that only 20 percent of a software program's capability is ever used, even with software as ubiquitous as an e-mail client. One reason is that the features do not relate to your firm's discipline. For example, you may never have any reason to use multilines or tolerancing symbols in your drawings.

Another reason for underutilization is not knowing about the features in the CAD system. This can be due to (1) insufficient training; (2) lack of initiative to try out unknown commands and explore the documentation; or (3) fear of the unknown — "I might break something." Let's look at each of these in turn.

Insufficient training. Overcome this with noon-hour tips sessions. I recall how a client reacted in amazement when I showed him the **Offset** and **Trim** commands; he had no idea they existed in AutoCAD. I suspect the cause is quite typical: you wish the software would do a task for you, but you have no idea how to find it, or even if the tool exists. Part of the problem is that today's software sometimes does too much; the other part of the problem is that the programmer's implementation (and naming) of a tool differs from your idea of it. For example, I had been using Word for five years before I discovered its **Change Case** command; I had never noticed it on the **Format** menu.

Lack of Initiative. There are those for whom CAD drafting is "just a job." They lack the initiative to find out how to perform tasks more efficiently; there is no ambition in finding out what else their CAD software can do; they have no interest in what's going on under the hood of the software and hardware. And this occurs in all industries. For example, I've asked the teller how his bank calculates the exchange rate on foreign currency; the teller has no idea: ""I just get it off the computer."

Fear of the Unknown. This last item is not an irrational fear; too often I have found my own computer operating more poorly when I tried something different. For example, around the time that I was writing this book, I upgraded my notebook computer to Microsoft's Internet Explorer v5.5. After the upgrade was complete, the File Manager took minutes to display the contents of a drive. I reversed the upgrade, which fixed the problem of the slow File Manager, but now connecting to the network took 15 minutes — something that previously occurred in a second or two.

CAD MANAGEMENT ISSUES

Not everything connected to CAD management involves explaining the corporate layering scheme to new employees and ordering additional plotter paper. At time, the CAD manager needs to spend some time thinking about broader issues.

The last part of this chapter includes a several brief essays covering some of the broad CAD issues that have been raised by readers of my *upFront.eZine* e-newsletter.

Do CAD Brands Matter Anymore?

The notion that CAD brands might matter less in the future came to me when I heard an interview with an author talking about the revolt against brand names. During the interview, a caller asked if brand names matter any more. He noted that the same factory in China makes several brands of running shoe — the name Nike, Rebook, or Adidas is glued on, and the shoe is shipped to the rest of the world. If, he asked, Nike no longer designs the shoe or makes the shoe, does it matter if the name of the shoe is Nike? It seemed to him that brand names no longer matter, because there was nothing behind the name.

The issue of brand name is very strong in the CAD world. AutoCAD. MicroStation. CADAM. MicroGDS. TurboCAD. Visual CADD. SolidWorks. Generic CADD. Solid Edge. ArchiCAD. VectorWorks. allPlan. IronCAD. Revit. IntelliCAD. DataCAD. CADKEY. XCAD. SmartSketch. IGDS. Pro/Engineer. Unigraphics...

Users are loyal, or resigned, to a single brand name. Loyalty comes from two areas:

- CAD is hard to learn. Users put a lot of sweat equity into learning CAD.
- CAD files are landlocked. Drawings are not easily exchanged with other CAD systems.

As a result, most users are familiar with a single CAD package, and tend to be unaware of the other brands. But how strong is the brand name of that CAD package? Less and less software is written by programmers employed by the CAD vendor. Instead, programming is farmed out to firms in India, Russia, and other locations where brilliant programmers are willing to work for lower-than-

---

"When writers give their readers exactly what they want, the readers are seldom enriched. They hear only what they already know; their prejudices are confirmed, their weaknesses pandered to. The audience is entertained, but not challenged or instructed."
— *Gene Edward Veith, Jr.*
  "Postmodern Times"

"Management launches a skunkworks because it has failed to create an organization that can be innovative without skunks."
— *Michael Schrage,*
  Fortune magazine

> "We feel that the technology industry is at this point [in 2001] where the technology has exceeded a lot of people's capacity to take advantage of it."
> — *Michael Tiemann, CTO, Red Hat Software*

North-American wages. A single programming house provides code to many CAD packages.

For example, CAD publications run ads from international programming firms that state, "Some of the major players in the CAD/CAM arena are our clients. We are able to develop quality software at a fraction of your own in-house development cost, using our India setup."

Reports one CAD magazine editor, "I had a prospective email from Serbia yesterday, offering programmers at $25 per hour to do pretty much anything. I think brands only matter when it comes to translating data." When people ask me which CAD package to buy, I say, "Buy the CAD package your clients use." Compatibility of the drawing files is more important than the brand name.

Author Jeremy Rifkin says: "The key is to find the appropriate mechanism to hold on to a consumer for life. Part of this is the leasing culture. Another part is to create communities of interest."

In CAD, we now see vendors now trying the leasing culture via a variety of subscription models, such as Bentley's SELECT (pay an annual fee and get a CD-ROM mailed to you every so often) and Revit (pay a monthly fee and get updates via the Internet). As for communities of interest in the CAD world, these would be portals, both independent (such as www.tenlinks.com) and vendor-sponsored (such as Autodesk's Point A).

After I published this editorial in *upFront.eZine*, readers generally reacted negatively. "This is a misdirected argument because it applies the concept of branding to an industry with highly functionally differentiated products, rather than an industry with negligibly differentiated products, such as running shoes," wrote one reader. Another wrote, "The brand does matter if the company behind the brand is still engaged in quality control and actively specifying the design to the production team." Yet another wrote, "Where a product is manufactured or who makes it is irrelevant to a brand; a brand mirrors the quality of the product and the company behind it."

> "Everyone wants to be like Nike. No company, no factories. It's a brand."
> — *Jerremy Rifkin*
> *"The Age of Access"*

While brand names may well survive, the CAD vendors themselves may not. The January, 2000 issue of *Engineering Automation Report* (www.eareport.com) predicts that today's many CAD vendors will consolidate. In the mechanical CAD field, the newsletter suggests that four vendors will survive in the long run: PTC, UGS, Dassault, and Autodesk.

The Case Against CAD Management

Volker Mueller is design technology manager at an architectural firm in Ohio, United States. He had several thoughts on CAD management after reading the following in my upFront.eZine e-newsletter #227: "Architects have always been called 'builders.' But architects don't build; they are in the information business. Unlike the old saying that "information is power', it is actually understanding information that is power."

CAD Management needs to disappear. We do not have "word processing management"; why should we need "CAD management?"

The issue, of course, is much bigger. As I remember it, the saying is "*Knowledge* is power." Knowledge goes beyond understanding of information. Along these lines, the issues of CAD management become issues of knowledge management. Suddenly we burst the convenient containers of expertise.

The more the push of information integration through the CAD environment has progressed, the further the CAD manager of old times has been left behind. The observation that "less than 1 percent of new buildings have a digital 3D model in the year 2000" is an intriguing indication of that fact. Back in the mid-1990s, the ratio of established use of technology (digital drafting) to outdated technology (manual drafting) would have been much higher (admittedly anybody's guess, but perhaps 70 percent?).

The single building model is not yet state-of-the-art, but will we have 70 percent saturation in five years? I doubt it, but I am willing to be pleasantly surprised.

Isn't Revit Technology Corporation's success (at least PR-wise, and with *Computer Graphics World* magazine's innovation award) built on the premise that we should not need CAD management? I really do believe that in those design firms that are somewhere on the leading edge, CAD managers will disappear by evolution into knowledge managers. Eventually, every contributor to a project should be able to do part of what the CAD manager does now.

Why not have a standards team that develops and maintains the order that currently CAD managers are supposed to enforce? Just because it is technically too difficult to be implemented by the average user? This is not good enough a reason.

There are people who think they know enough about all the issues in knowledge management because they are able to surf the Internet.

In a newsletter, mention was made of the development of CAD systems into information browsers (graphic, alphanumeric). We

"You drop your calculator and the battery falls out, and suddenly you haven't got a brain anymore."
— *Brian Winterflood, Winterflood Securities*

"Accumulation of wealth has no purpose."
— *Jim Barksale, Former CEO Netscape*

> "Because computer source code is an expressive means for the exchange of information and ideas about computer programming, we hold that it is protected by the First Amendment."
> — Unanimous ruling of an American three-judge appeals court panel.

should be able to get to documents that are referring to a certain symbol in a drawing — a specific object in a model — through the drawing. The tedium with which that currently happens needs to go away. I should be able to drag a product's spec document onto the graphic representation of the product, so that later, when selecting the product, I am able to retrieve the spec document without additional effort. Maybe one of those CAD packages out there allows me to do that already; it just is not the one our firm uses.

Not only are CAD managers being left behind. The same fate applies to project architects, project managers, and other people who felt that they could stay comfortably detached from the day-to-day activity of CAD. Rude awakening ahead! They will not have a clue about how to access the information that they need to run the show. Even worse, they will not know how to generate it, and how to assemble it.

Suddenly the CAD managers not only need to become the knowledge managers, but they also need to become the knowledge management trainers. There are plenty of years of work ahead in that area. Which CAD manager is up to that challenge?

A question: how much will this group of people drive the future development of the industry? Answer: they won't. But then again, not everybody needs to innovate. The CAD managers who do not evolve will be left behind as fossils and will continue to serve fossil firms well.

The Disadvantages of CAD

A student at a British university asked readers of upFront.eZine for help. He had found it easy to obtain information about the benefits of CAD, but could not find any statements on the disadvantages of CAD over manual drafting. Here are some responses:

- Less spontaneity compared to scribbling with pencil and paper.
- Increased health problems, such as carpel tunnel syndrome, back and shoulder pain, and less moving around.
- Lack of adherence to standards makes it more difficult a CAD operator to take over another's digital drawings.
- Difficult to see the entire drawing until it has been plotted; difficulty of making minor, last minute changes.
- Dependence on electricity; more paper wasted.

> "Nine times out of ten, there's no compelling reason to upgrade. Whatever wonderful features version x.0 offers must be weighed against all of the flaws that new software inevitably contains."
> — Licoln Spector
> PC World magazine.

Is CAD Dead?

Martyn Day is the editor of several British CAD publications, including MCAD and CADdesk AEC. As one of the most outspoken CAD editors, his question comes as no surprise.

Another day, another corporate re-branding. I am hard pressed to find the word "CAD" on any of the home pages of traditional CAD vendors. Instead, you find a new breed of three-letter acronym: Bentley emphasizes EEM (Enterprise Engineering Modeling); PTC is the collaborative product commerce (CPC) company; Autodesk recently announced its future roadmap is idesign.

"CAD" has been a dirty word for the last several years in the stock market. The reason was a double-whammy: the rise of Internet tech stocks, combined with the CAD market maturing and offering less opportunity for growth. For the stock market to like a tech stock, your software house had to be producing a revenue growth in excess of 20% per year. This was consistently achievable for a number of the key players, like Autodesk and PTC.

With CAD sales looking unlikely to enjoy the growth the stock market would reward, CAD companies are diversifying into developing other tools with which to leverage the engineering documents created by their design software solutions. The area of greatest interest is electronic document management (EDM), where each player already has incumbent installed bases and the sale was to the whole enterprise, not just the CAD department.

In this era of "if you can touch it, it's worth nothing," the companies that provide the tools for the manufacturing industry are busy remodeling their businesses along Internet lines so at least they can improve their market capitalization. To back this up, the marketing departments have clearly been burning the midnight oil, reinventing the direction and rejuvenating the images of these apparently tired old CAD companies.

I'm pretty sure I'm not the only person who feels this way. I have yet to talk to a marketing contact that doesn't at least manage a snigger when I mention the roughage-quality of their latest Net-centric press release. They know it's bobbins, I know it's bobbins.

In short, Internet re-branding has become a joke within our industry. It transpires that the target audience for this posturing, however, is the tiny stock analyst community — the folks with the power to mark stock as Buy, Hold, or Sell.

Above all else, expect to see more Internet functionality in your design software, whether you need it or not.

"**Q:** What are you planning to do about all the different kinds of Linux around? It's so confusing out there!"
A: People from East Germany have found the West so confusing. It's so much easier when you have only one party."
— *Linus Torvold founder of Linux*

"Theory is when you know something, but it doesn't work. Practice is when something works, but you don't know why. Programmers combine theory and practice. Nothing works, and they don't know why!"
— *Boston Computer Society*

The Travails of Archiving

Shane Beaman of Sealcorp Computer Products, New Zealand, began wondering about how to deal with drawings that are getting old. The problem of how to archive paper drawings formerly dominated; but now with CAD getting to be more than 20 years old, archiving digital drawings is becoming an issue.

Has anyone ever looked seriously at electronic drawing archiving and recovery? I want to be able to access design data created and stored electronically 15 or more years ago, or archived now for access in future years.

How do I get information off old hardware (8" and $5\frac{1}{4}$" floppy diskettes, $\frac{1}{2}$" and $\frac{1}{4}$" tapes, etc.) stored on old design software that is possibly no longer readable by today's software? I guess I would have to keep a copy of the old design software, but I wonder if it will still be readable, or able to be run on the new hardware.

Everyone today is encouraged to store information electronically because it is cheap, reasonably secure from corruption, and large amounts of data can easily be stored. But what happens in 20 to 30 to 40 years time when I need to read or retrieve a design that was created on old software or hardware?

Imagine a consulting firm that has created hundreds of drawings all over the country every day and stores them for many years. When there is a new release of software or hardware, do they retrieve all the old designs, and convert them onto the new hardware or the latest software? I don't think so.

What happens when a software vendor discontinues a design product or goes out of business, and I need read some old drawings that were created years earlier and I no longer have the software to read them? Says something for sticking with a large vendor who'll likely be around in some form in years to come. Also says a lot for keeping printed and bulky documentation on file.

Even if we had a universal data format, we still have the problem of being able to read the drawings off old hardware. With the way the computer industry changes, who knows if we will have the same data storage devices or format devices that can be read in 20 years.

Readers of *upFront.eZine* gave the same general response: "Yup, ya gotta update those archives with each turn of the technological screw." Wrote one reader, "That is exactly what must be done to make CAD data useable. Over a 10-year period, our archives have

---

"It doesn't start from trying to figure out what people want, then giving it to them. I think it should start from me doing something that I'm excited about."
— Joe Jackson
musician

"The real topic in astronomy is the Cosmos, not telescopes. The real topic in computing is the Cybersphere and the cyberstructures in it, not the computers we use as telescopes."
— David Gelernter
The Edge

existed in three different locations and two different CAD formats." Another reader wrote, "Bring really important data, such as coordinate data, forward. But at most, leave data one step back."

A software vendor suggested, "Those who archive paper drawings scan and save them as raster files onto CD or hard disk. Unlike proprietary CAD file formats, the structure of TIFF, CALS, BMP, etc, raster files are all in the public domain and will be accessible in 20, 30, or 40 years time to anyone willing to decode the format."

One reader said his firm made the decision to abandon old data. "We made this decision for several reasons: (1) cost and space of maintaining old hardware; (2) converting data requires significant time and does not produce a great product; (3) CAD operators said they preferred to recreate clean data, rather than deal with the mess produced by the port."

It's relatively easy to bring old 2D CAD data forward to modern standards. Another reader raised the problem of modern CAD formats. "We don't think we can store CAD drawings containing 3D, paper space views, and intelligent objects — it will be another problem when it becomes to read such data."

And several reminded us to not store data magnetically. Avoid diskettes, ZIP drives, and MO (magneto-optical) disks because the magnetic flux weakens with time. Use CD-R and DVD discs, and keep the media fresh.

Don Beaton, the technical editor for this book, made the following comments on archiving of CAD data:

The discussion of longevity of CAD data needs to be expanded to include issues such as hardware standards, operating systems, hardware drivers, hardware interface boards and cables. For example, as this book is being written, Compaq says it will start shipping desktop computers without parallel and serial ports.

There are so many different forms of backup media, all incompatible with each other: magnetic tape (4mm DAT, 8mm Exabyte, 9-track, Colorado), floppy (8", $5\frac{1}{4}$", $3\frac{1}{2}$"), punched paper tape, CD-ROM, DVD-RAM, ZIP disks. In addition, a particular piece of hardware, such as a tape drive, sometimes cannot read tapes created on an identical drive due to problems like head-alignment error. Backup media have other problems, too. There is the destruction of magnetic media due to:

> "Generally plainspoken, few contractors appreciate the unrelenting hype that construction industry dot.coms emit daily through ads, press releases, and marketing material, each claiming to be the greatest."
> — *Engineering News Record*

> "Why should these choices be Microsoft's choices, and why is it getting even more difficult to make alternative choices?"
> — *The Register commenting on Microsoft's plan to hardwire Internet Explorer v6 with Windows Media Player, Instant Messenger, My Pictures, and extensions to DHTML.*

- Elevated temperature: in a car or a corner office on a sunny day.
- Fire.
- Water damage: flooding or sprinkler systems putting out that fire.
- Magnetic fields: telephone, purse latch, paper clip storage container, fridge magnets, static from carpets, field around wires.
- Tape "eaten" by the tape drive.
- The magnetic pattern on the tape can alter the pattern on adjacent tape if it is not rewound occasionally.
- Human error: a valuable tape is discarded, or is reused for new backup.

Never discard old backup hardware: you may need it to restore data from obsolete media. The following Web pages discuss the long-term view of data storage:
www.longnow.com/10klibrary/library.htm
www.sciam.com/0397issue/0397lesk.html

Consideration needs to be given to publishing CAD drawings in formats likely to be viewable in the long term. That means saving the files in a nonproprietary format that can be viewed with free, open source viewers. Almost all CAD formats, including DWG, are proprietary. To be considered nonproprietary, the file format must meet two conditions: (1) the file format must be published; and (2) changes are made to the format by a committee.

One storage format currently popular is Adobe's PDF. The advantages include: file cannot be altered; file is self contained (no external font files, for example, are needed); fonts and xrefs are displayed correctly; details can be plotted; format is vector, so it looks good when plotted at various sizes; colleagues can make annotations and comments, which are stored on a separate layer from the data; and the Acrobat reader is free.

Some CAD packages can create a PDF file directly. If your's cannot, from add a PostScript plotter (such as the HP 755cm), and plot the drawing to a PS file. Convert the PS file to PDF format.

"No one ever went broke giving people what they want. Nope, you go broke charging more than people want to pay for a product they can get elsewhere."
— Carlton Dow,
  The Peddie Report

"I've heard umpteen stories about clueless consultants and marketing newbies who wonder aloud at meetings why advertorials [editorial written by the advertiser] need to be labeled, why paid links need to be disclosed, why promotions needs to be treated differently than editorial contents."
— J. D. Lasica
  Online Journalism Review

Stay or Switch?

Every so often, a CAD vendor hopes to increase their market share by enticing customers from competitors. While these campaigns convince some users to switch (perhaps those who were planning to anyhow), the efforts generally fail and the CAD vendor gives up. At least, until the marketing department dreams up another campaign.

CAD vendors are always looking for ways to siphon off users who use a different CAD package. In particular, AutoCAD users are the primary target. One enticement is to offer a competitive upgrade offer. For example, SofTech offerer a discount to expand its market share of its CADRA software. Regularly $3,995, the software was offered at a 45 percent discount for a limited time when it "replaces industry rival Autodesk Inc.'s AutoCAD seat, the unit price is $2,199."

Sometimes, the urge to switch comes via advertising, offering a better product. A three-page ad from SolidWorks ran for several months in CAD magazines telling AutoCAD users, "Unlike Mechanical Desktop, which was built on a 15-year-old 2D foundation, SolidWorks software was designed from the ground up as a 3D tool."

Other times, the CAD vendor will create a free set of migration tools. These tools typically include a one-way file translator, a tutorial on using the new CAD software, and documentation helpful to making the transition.

I asked the readers of *upFront.eZine* if they would you actually pay $2,199, or $495, or even nothing to try out a different CAD package. Their response surprised me by being balanced: roughly equal numbers were against, in favor, or didn't care.

Wrote one reader, "To switch in a business from a know, proven program with reasonable compatibility and support to an unknown product is not a prudent decision. There is the cost of retraining, lost time, and new add-on programs to purchase and learn. A thousand dollars does not near cover these problems." Another wrote, "New users will relish the abundance of CAD software and undoubtedly select the right tool for their application."

"Primarily, these financial offers from CAD companies are successful with users of illegal licenses who see this as a way to get legal at a much lower cost," wrote a reader. Another wrote, "We

---

"The size of a piece of electronic equipment can be used to date it, not unlike the number of rings in the cross-section of a tree."
— *Jason O'Grady*
Go2Mac.com

"How much cheese do you take off a pizza until you have no customer left?"
— *Gordon Bethune*
CEO Continental Airlines

> "It's the kind of gee-whiz 'news' we'd like to see less of: a trade association or professional group sponsors research that says its business sector is going gangbusters, then deadline-crazed business journalists jump on the findings and dutifully spit 'em out with little analysis of their own."
> — Media Grok
> *The Industry Standard*

often get offers from other CAD architecture vendors, who made us offers to change to them, but we never changed to them because we really never had the time to change. It would take too much time to learn another cad system."

On the other hand, a number of readers reported that they might switch (under the right conditions), had tried switching (but went back), or successfully switched. "As a long time user of AutoCAD, I was surprised to find how easy it was to switch. For 2D I am now using IntelliCAD, which I find slightly better than AutoCAD and it fits my new consulting business budget," wrote one reader, while another downgraded from AutoCAD: "AutoCAD LT has replaced AutoCAD for us."

A reader wrote, "I made three or four attempts to upgrade from Generic CADD to VisualCADD, and this is the same product line. I can't imagine what switching to an all together different system would be like." And another reader reported, "I tried other CAD programs, only to go back to AutoCAD." Sometimes, it's the customizing that keeps you with a CAD package: "We have over a quarter-million lines of LISP that would have to be rewritten in VB or VBA, and that ain't gonna be cheap."

Are Vendors Helping CAD Managers?

As you will read in the coming chapters, CAD vendors seem reluctant to include standards with their software package. Only a very few, for example, include the AIA's system of naming CAD layers.

The emphasis among CAD vendors seems to be on increasing the stock price by making Wall Street happy. This means, in some cases, adding jingoistic features, such as the orgy of Web-oriented features during the year 2000. (Many of the features added to the CAD programs are equally accessibly through any Web browser.)

Toward the end of 2000, Autodesk released an add-on for managing CAD. The add-on didn't provide standards, but rather reviewed drawing files for compliance with standards, in either an interactive or a batch mode (handy for overnight processing of large numbers of drawings). The problem was that the add-on software (1) worked only with the 2000i release of AutoCAD; and (2) cost $149 per copy.

> "When you know something is wrong and you don't challenge it, you have become part of the bureaucracy."
> — Arthur Black
> *Home Depot cofounder*

Other functions that would help the CAD manager also incur extra cost — drawing management systems; drawing viewers and redlining; project management; published standards from semi-official bodies; training; and so on. CAD vendors are content to throw in flashy extras for free — such as photorealistic rendering and 3D solids modeling, which perhaps help make software sales — but include very few functions useful for the management of computer aided design.

> "Truly great companies aren't built by the greedy, but by the passionate."
> — *J. William Gurley*
> *Above the Crowd*

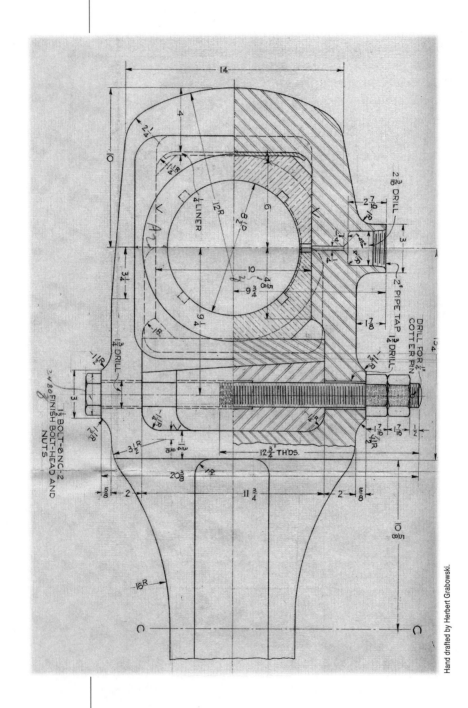

Hand drafted by Herbert Grabowski.

Naming Drawings and Creating Symbols

2

Drawings must be given file names that allow you to identify them by project and discipline. Folders (subdirectories) are also used to segregate projects.

The file name of a drawing is closely related to the sheet number. If you decide to use a CAD drawing to produce a single plotted sheet, then the drawing's file name is the same as the sheet number. If, however, a CAD drawing produces several different plotted sheets (either via toggling the visibility of layers, or via referenced drawings) — which is how CAD ought to be used — then the sheet number must differ from the drawing file name.

Here are several examples of drawing file name conventions:

Simple Filename Convention

The simplest file name convention is based on the project number. If necessary, it can be further refined by discipline, and so on. For example, a drawing with the name 60591A01.DWG has the following meaning:

60591	Project number
A	Discipline (A = architectural)
01	Drawing number
DWG	The AutoCAD file extension

Digits and letters are alternated to help distinguish the parts of the file name.

In the next example, the file name A-P-06-915-B.DGN uses dashes to separate the parts of

CHAPTER SUMMARY
In some CAD packages, symbols and files are closely related; thus, we look at both in this chapter. For example, a symbol external to AutoCAD is simply a drawing file. In this chapter, you learn how to:
· Set up file names and folders for your drawings.
· Create standards for symbols.

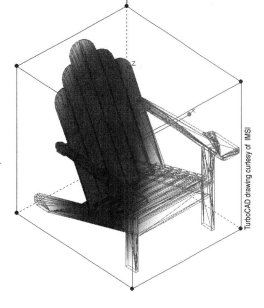

TurboCAD drawing curtesy of IMSI

the file name. In addition, the file name allows for revision numbers:

A	Discipline
P	Drawing type (P = plan drawing)
06	Detail number
915	Sheet number
B	Revision number
DGN	MicroStation file extension

The next example of a drawing file name, HQB2E001.RVT, allows for floor numbers for a building project:

TS	Drawing description (TS = tool shed)
B2	Floor number (B2 = second basement floor)
E	Discipline (E = electrical)
001	Sheet number
RVT	Revit file extension

Here are several examples of drawing file names for a large roadway design project using externally referenced files:

RW01	Right-of-way requirements, sheet 1
PL03	Plan, sheet 3
GM05	Road geometry, sheet 5
PM07	Pavement markings, sheet 7

Industry Standards for File Names

The following sections describe file name standards created by several industry groups:

- American Institute of Architects
- Construction Specifications Institute
- California Department of Transportation

AIA File Naming Convention

The American Institute of Architects uses a file naming system based on its *CAD Layer Guidelines* (see Chapter 3 of this book). The naming convention recognizes two types of CAD files:

- **Model files**: the model drawn full size, but plotted to scale.
- **Sheet files**: non-model parts of the drawing, such as dimensions and border; plotted full size.

Model file names begins with the *agent-responsible* code (such as A for Architect), followed by a two-letter drawing type, followed by a sequential number. For example, A-FP-0001.dwg is the first drawing of the architectural floor plan.

Sheet file names use numerical descriptors.

Code	Agent Responsible
A	Architect
C	Civil
E	Electrical
F	Fire Protection
I	Interiors
M	Mechanical
P	Plumbing
S	Structural
T	Telecommunications

The drawing type code applies to all disciplines:

Code	Drawing Type
3D	Isometric/3D
DG	Diagrams
DP	Demolition Plan
DT	Detail
EL	Elevation
FP	Floor Plan
QP	Equipment Plan
SC	Section
SH	Schedules
SP	Site Plan
VP	Evacuation Plan
XP	Existing Plan

DISCIPLINE CODES

Here is a list of discipline code letters you can employ for drawing file names:

- **A** Architectural plan
- **C** Civil and site plan
- **D** Demolition plan
- **E** Electrical plan
- **F** Food services plan
- **G** Graphics and signage plan
- **I** Interior design
- **L** Landscaping plan
- **M** Mechanical plan
- **P** Plumbing plan
- **S** Structural plan
- **T** Tenant plan
- **U** User-defined plans

And here's a list of drawing type codes letters for file names:

- **B** Blocks (symbols) and external references
- **C** Composite drawings
- **D** Detail drawings
- **E** Elevation drawings
- **N** Enlarged plans
- **P** Plans
- **S** Section drawings
- **T** Text schedules
- **W** Wall sections

In addition, there are drawing code types specific to each discipline:

Discipline Codes	Meaning
Architecture	
A-CP	Ceiling Plans
A-EP	Enlarged Plans
A-NP	Finish Plans
A-RP	Furniture Plans
Civil	
C-EP	Environmental
C-GP	Grading
C-RP	Roads/topography
C-SV	Survey
C-UP	Utility
Electrical	
E-CP	Communication
E-GP	Grounding
E-LP	Lighting
E-PP	Power
Fire Protection	
F-KP	Sprinkler Plan
Interior	
I-CP	Ceiling Plans
I-EP	Enlarged Plans
I-NP	Finish Plans
I-RP	Furniture Plans
Mechanical	
M-CP	Control Plans
M-HP	HVAC Ductwork Plans
M-PP	Piping Plans
Plumbing	
P-PP	Plumbing Plan
Structural	
S-FP	Framing Plans
S-NP	Foundation Plans
Telecommmunications	
T-DP	Data
T-TP	Telephone

CSI Uniform Drawing System

In 1994, the Construction Specifications Institute created a drawing sheet naming system that, like the AIA's, is based on the CSI's layering system. The drawing file names are based on the order of a construction project. Called USD (uniform drawing system), the system uses two characters, followed by three digits, as in AS102.

Level	Meaning
A	Level 1: Discipline designator (A = architecture)
S	Level 2: Discipline modifier (S = site)
1	Level 3: Sheet type (ranges from 0 to 9)
02	Level 4: Sequence numbers (ranges from 00 to 99)

To make it easier to remember, UDS uses the first letter of a discipline as the Level 1 designator, with a few exceptions:

Level 1	Discipline Designators
G	General
H	Hazardous Materials
B	Geo technical
V	Survey/Mapping
W	Civil work
C	Civil
L	Landscape
S	Structural
A	Architectural
I	Interiors
Q	Equipment
F	Fire Protection
P	Plumbing
M	Mechanical
E	Electrical
T	Telecommunications
D	Process
R	Resource
X	Other Disciplines
O	Operations
Z	Contractor/Shop Drawings

If a modifier is not used, a hyphen is used as a placeholder.

OTHER UDS SPECS

In addition to organizing drawings by file name (module 01), the CSI's Uniform Drawing System also specifies (module # in boldface):

02 Sheet Organization
03 Schedules
04 Drafting Conventions
05 Terms and Abbreviations
06 Symbols
07 Notations
08 Code Conventions

Naming Drawings and Creating Symbols

Level 2	Discipline Modifiers	Level 2	Discipline Modifiers
General		**Interiors**	
GI	Informational	ID	Demolition
GC	Contractual	IN	Design
GR	Resource	IF	Furnishings
		IG	Graphics
Hazardous Materials			
HA	Asbestos	**Equipment**	
HC	Chemicals	QA	Athletic
HL	Lead	QB	Bank
HP	PCB	QC	Dry Cleaning
HR	Refrigerants	QD	Detention
		QE	Educational
Civil		QF	Food Service
CD	Demolition	QH	Hospital
CS	Survey	QL	Laboratory
CG	Grading	QM	Maintenance
CP	Paving	QP	Parking Lot
CI	Improvements	QR	Retail
CT	Transportation	QS	Site
CU	Utilities	QT	Theatrical
		QV	Video/Photographic
Landscape		QY	Security
LD	Demolition		
LI	Irrigation	**Fire Protection**	
LP	Planting	FA	Fire Detection and Alarm
		FX	Fire Suppression
Structural			
SD	Demolition	**Plumbing**	
SS	Site	PS	Plumbing Site
SB	Substructure	PD	Process/Plumbing Demolition
SF	Framing	PP	Process Piping
		PQ	Process Systems
Architectural		PE	Process Electrical
AS	Site	PI	Process Instrumentation
AD	Demolition	PL	Plumbing
AE	Elements		
AI	Interiors		
AF	Finishes		
AG	Graphics		

Level 2	Discipline Modifiers
Mechanical	
MS	Site
MD	Demolition
MH	HVAC
MP	Piping
MI	Instrumentation
Electical	
ES	Site
ED	Demolition
EP	Power
EL	Lighting
EI	Instrumentation
ET	Telecommunications
EY	Auxiliary Systems
Telecommunications	
TA	Audio Visual
TC	Clock and Program
TI	Intercom
TM	Monitoring
TN	Data Networks
TT	Telephone
TY	Security
Resource	
RC	Civil
RS	Structural
RA	Architectural
RM	Mechanical
RE	Electrical

CalTrans Filenaming Convention

The California Department of Transportation uses three naming conventions for design files: project, related, and topographic drawings.

The file name of project drawings has the following format: D12345E01.DGN

D	District code
12345	First five digits of the project expenditure authorization
E	Sheet identification code (E = electrical)
01	Sheet number; letters A through Z represent sheet numbers 100 through 125.

The file name of related drawings has the following format: DRHGKHSHR.DGN

D	District code
RHG	Operator's initials
KHSHR	Drawing name assigned by the operator

The file name of topographic drawings has the following format: D86725B27.DGN

D	District code
86	Fiscal year of the HQ aerial survey contract
725	Contract order number
B	Drawing code (B = base map)
27	Sheet number; sheet numbers 100 and higher eliminate the drawing code

CalTrans uses code letters, shown in the following table, to identify the type of drawing in the file name:

Drawing Type	Plan	Sheet ID
Sign plan	S	P
Retaining wall	R	Q
Sound wall	SW	R
Roadside rest	...	S
Planting and irrigation plan	HP	T
Lighting and signal plan	E	U
Revised standard plan sheets	...	V
Cross sections	Z	Z
Master design map	...	AA
Topographic base map	...	BB
3D terrain data	...	3D
Scanned drawing	...	CC
Digitized drawing	...	DD
Created drawing	...	EE
Project file directory	...	FF
Route adoption map	...	GG
Area of interest map	...	HH
Strip map	...	II
Freeway agreement map	...	JJ
New connection report exhibit	...	KK
PUC exhibit	...	LL
Geometric approval drawing	...	MM
Bridge site map	...	NN

FILE EXTENSIONS

The three-letter extension is used by the CAD software to identify its own file format. Some examples are shown in the following table:

File Extension	CAD Package	Meaning
2D	VersaCAD	2 Dimensional
3DM	Rhino	3D modeler
3DS	3D Studio	3D Studio
ATX	Actrix	AcTriX
BMF	FloorPlan 3D	Building
DC5	DataCAD	DataCad, v5
DGN	MicroStation	DesiGN
DW2	DesignCAD	DesignCAD Workfile, v2
DWG	AutoCAD	DraWinG
FCD	FastCAD	FastCaD
IGR	SmartSketch	InterGRaph
MCC	Monu-CAD	Monu-Cad Cad file
PAR	SolidEdge	PARt
PL1	Chief Architect	Plan, 1st floor
PLN	ArchiCAD	PLaN
PRT	CADkey	PaRT
PRT	SolidWorks	PaRT
PRT	Unigraphics	PaRT
RVT	Revit	ReViT
SDP	CoCreate	Solid Designer Part
SDP	SolidDesigner	Solid Designer Parts
SKD	AutoSketch	SKetch Drawing
SOL	Matra	SOLids
T2T	Sonata	unk
TCW	TurboCAD	TurboCad for Windows
VCD	Visual CADD	Visual Cadd Drawing
VSD	Visio	ViSio Diagram
VWF	Cadvance	VieW File

RESOURCE

A long list of CAD file extensions is available at www.cad2cam.com/list/ , while the following site lists general computer file extensions www.crosswinds.net/san-marino/~jom/filex/extensio.htm

WHAT ARE SYMBOLS?

Drafting often involves drawing many similar symbols. Whether windows, valves, or transistors, symbols are drawn many times over. Back in the days of manual drafting, there were two common solutions: trace the pencil around a green plastic template, or stick see-through photocopies onto the Mylar.

With CAD software, there is just one solution: insert a previously drawn symbol. A single command — even just a drag'n drop action — places complex symbols without drawing a single line! The following figure shows an electrical drawing made with TurboCAD. The drawing was created by inserting symbols, then joining the symbols with lines and adding text.

In CAD, the rule of thumb is: **Anything drawn twice should be turned into a symbol.**

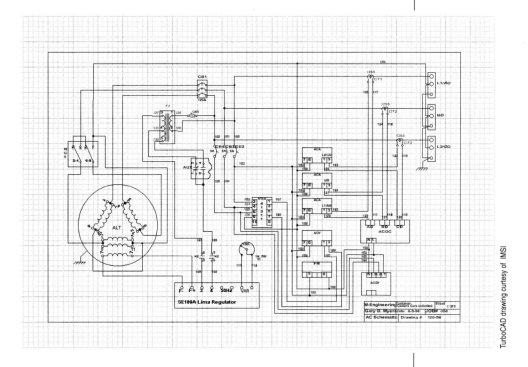

With the symbol placed in a drawing, you'll find it easier to edit symbols than individual entities. A single pick selects the symbol for editing, rather than picking all individual entities. When you update a symbol, all copies are updated.

SYMBOL CREATION SUMMARY

Follow these nine steps to create symbols for CAD drawings:

1. Draw the symbol.
2. Decide its orientation.
3. Select an insertion point.
4. Determine the scale: full size or unit size.
5. Appoint layer 0 or a specific layer.
6. Add attribute (tag) information.
7. Name the symbol based on the CAD package's naming limitations.
8. Document the symbol in a three-ring binder or electronically.
9. Store the symbol in a read-only folder; keep a backup copy offsite.

Symbols in CAD

Most CAD systems allow you to place an unlimited number of symbols in a drawing; you repeat each symbol as often as required. Employing symbols in CAD drawings is more efficient (uses less memory) because multiple occurrences of a symbol are referenced. If you insert a bathtub symbol three times, the drawing file contains one copy of the bathtub and three references to the symbol definition.

How to Create a Symbol Library

To make your firm's drafting more efficient, create a central library of symbols to which all CAD operators have access. There are nine steps to creating a symbol:

Step 1. Draw the Symbol. When drawing the symbol, recognize that there are three basic types, which means they are each drawn somewhat differently (see figure):

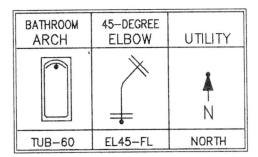

A CAD symbol is either literal (left), symbolic (center), or a reference (right).

- **Literal symbols** imitate physical objects, such as bathtubs and plants.
- **Symbolic symbols** that represent physical objects in iconic form, such as a light switch and a piping elbow.
- **Reference symbols** indicate the location of nonphysical objects, such as the North arrow and scale bars.

Step 2. Decide on the Orientation. Symbols are usually placed in the drawing at different angles. Sometimes, a symbol is placed at one particular angle more often than any other angle. For example, the North arrow symbol is more likely to point to the top of the drawing; thus, draw it that way.

Step 3. Select the Insertion Point. The insertion point (a.k.a. reference point) is the point in the drawing where the symbol is placed. The symbol is usually placed at the pick point (a.k.a. data point). Like orientation, selecting an optimal insertion point now saves time later, although some CAD systems allow you to move the insertion point during symbol placement.

For example, the North arrow symbol would have its insertion point at its tip, to let you easily point it in the right direction. A bathtub symbol, on the other hand, would have its insertion point where the drain attaches to the pipe. For symbols with more than one logical insertion point, such as a piping symbol (which could have the insertion point at either end), select one end as the standard.

Step 4. Determine the Scale. The scale depends on the type of symbols. Literal symbols, such as bathtubs and plants, are drawn full size. For example, a five-foot bathtub is drawn 60" long. When the symbol is placed in the drawing, you specify a scale factor of 1.

Symbolic symbols and reference symbols are drawn to unit size. For example, a pipe elbow is drawn to proportionally fit a 1" square. Later, when you place the symbol, you scale the symbol to size.

Parametric symbols are different yet again. This is a literal symbol that is scaled to size during placement. For example, take a desk. In some CAD systems, you would save a single desk symbol as a 1" square. Upon placing the desk, specify scale factors of:

> x = 24" and y = 48" for a 24" x 48" desk
> x = 24" and y = 60" for a 24" x 60" desk
> x = 30" and y = 60" for a 30" x 60" desk

In other CAD systems, the dimensions of the desk are controlled by formulae. By using parametric formulae, a single desk symbol works for standard desk sizes. When working with large symbol libraries, parametrics save a lot of disk space.

> **TIP PHRASEOLOGY**
>
> Symbols are known by a wide variety of names:
> **Blocks** in AutoCAD
> **Cells** in MicroStation
> **Parts** in AutoSketch
> **Components** in Generic CADD
> **Groups** in Design CAD
> **Symbols** in Cadvance and TurboCAD

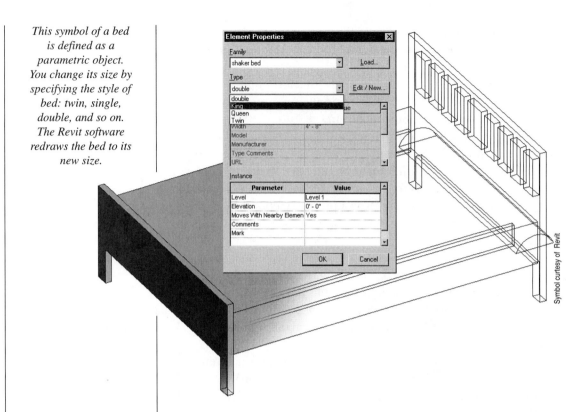

This symbol of a bed is defined as a parametric object. You change its size by specifying the style of bed: twin, single, double, and so on. The Revit software redraws the bed to its new size.

Step 5. Appoint the Layers. It matters which layer the symbol is created on. There are two kinds of layers on which to create the symbol:

- **Layer 0.** In some CAD packages such as AutoCAD, a symbol created on layer 0 has special meaning: when the symbol is placed in the drawing, it ends up on the current layer.
- **Any Other Layer.** A symbol created on any other layer is fixed to that layer: when the symbol is placed in the drawing, it ends up on the layer it was created and not the current layer.

For this reason, you must decide on your firm's layer system before you design the symbol library (see Chapter 3 of this book).

Step 6. Add Attribute Information. In some CAD packages, attribute data (a.k.a. tags) can be attached to symbols only. Attribute data is information that is extracted from the drawing and exported to a database or spreadsheet program. For example, the bathtub symbol might include attribute data that describes the manufacturer, model number, and color. In some CAD packages, you can link the symbol with a row in an external database table.

Step 7. Name the Symbol. Finally, decide on the symbol's name. Just as you must decide a standard for naming layers, you must create a standard for drawing and naming symbols. Most CAD systems now allow 255 characters in the name, but there is still a tendency to restrict symbol names to eight characters for compatibility and historical reasons. (Versions of MicroStation up to "J" limit cell names to six characters, but let you attach a long description.) You can apply some of the layer naming procedures (discussed in the next chapter) to naming symbols. In general, you may want to use a three-part name, as follows:

- **Discipline Name.** Reserve the first two characters of the symbol's name for the discipline. Here are some letter pairs to consider, based on the system used in Autodesk's old AEC package:

Discipline Code	Meaning
AR	Architectural
CD	Construction document
EE	Electrical
EQ	Equipment
FU	Furniture
LT	Lighting
MH	Mechanical, HVAC
MP	Mechanical, plumbing
TB	Title block
WS	Workstations

- **Part Name.** The second pair of letters describe the part, such as NC for a notebook computer and LS for a light switch.

- **Increment Number.** Use the last two or four digits to define different models of the same part, such as WS-NC-75 for a Model 75 notebook computer, and WWLS0203 for a two-pole three-way light switch.

RESOURCES

NIBS Facilities Information Council
National Institute of Building Sciences
1090 Vermont Ave NW
Suite 700
Washington DC
20005-4905 USA
www.nationalcad
standard.org

CAD Layer Guidelines
The American Institute of Architects Press
1735 New York Ave NW
Washington DC
20006 USA
www.aiaonline.com

The Uniform Drawing System
The Construction Specifications Institute
601 Madison St
Alexandria VA
22314-1791USA
www.csinet.org

Other symbol naming conventions you may want to consider include using the manufacturer's part number or adapting the CSI 16-Division system.

Step 8. Document the Symbol. After creating the symbol, document it. Many firms use a three-ring binder to document the symbol; others use electronic symbol librarians. Some firms document one symbol per page; others document four or nine symbols per page. Either way, here is the information you should include with each symbol:

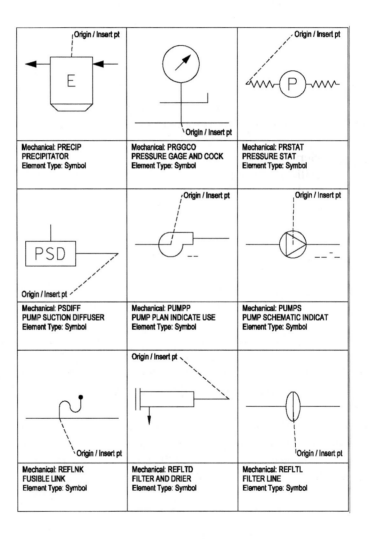

The Tri-Service CADD/GIS Technology Center provides well-documented symbols: the symbol is illustrated, named, and its origin is indicated.

- Symbol name
- File name (if different)
- Library file name (if on a CAD system that stores symbols in a library file)
- Description
- Insertion (or origin) point
- Orientation (if different from default)
- Layer created on
- Layer to be placed on (if different from creation layer)
- Default plot scale (full size or unit size)
- Attribute data (if any)
- Drawn by
- Last modified date
- If a nested symbol, describe constituent symbol

Step 9. Store the Symbol. When you are done creating and documenting the symbol, make the collection of symbols available to everyone in your office. If your office is networked, place all symbols in a single read-only folder, such as \CadSymbols.

The Tri-Service CADD/GIS Technology Center's *A/E/C CADD Standard* specifies the following directory structure for storing AutoCAD (Acad) and MicroStation (Ustn) symbols:

```
\Symbols
    \Ustn                               \Acad
        TsAEC.Rsc                           \Genl
        Genl_Sym.Cel                            *.DWG
        Surv_Sym.Cel                        \Surv
        Htrw_Sym.Cel                        \Htrw
        Cvls_Sym.Cel                        \Cvls
        Geot_Sym.Cel                        \Geot
        Util_Sym.Cel                        \Util
        Land_Sym.Cel                        \Land
        Stru_Sym.Cel                        \Stru
        Arch_Sym.Cel                        \Arch
        Intr_Sym.Cel                        \Intr
        Secu_Sym.Cel                        \Secu
        Fire_Sym.Cel                        \Fire
        Plmb_Sym.Cel                        \Plmb
        Mech_Sym.Cel                        \Mech
        Elec_Sym.Cel                        \Elec
        Tele_Sym.Cel                        \Tele
```

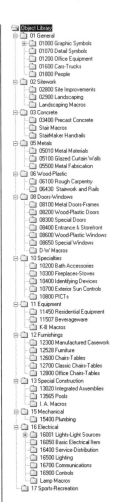

MicroStation stores its symbols (a.k.a. cells) in CEL library files. It, therefore, needs just a single folder in which to store all cells specific to a discipline.

AutoCAD, on the other hand, does not have symbol (a.k.a. block) libraries, so every block requires its own DWG files. To segregate blocks into disciplines, an additional levels of folders is needed.

Finally, make two backups of the symbol library, and store one offsite.

Sources of Symbols

Frequently, symbols are available from other sources, saving your firm the work of creating its own. A number of manufacturers make their product catalog available on diskette at no charge. Usually, the products are saved in DXF format, often with attribute information included. Since symbols are very simple drawings, no data is lost translating from DXF to your CAD system's format. One huge library of a million parts drawings from 7,000 vendors is available on-line and CD-ROM from Thomas Register at www.thomasregister.com

Some CAD vendors include a large symbol library with their product. Graphisoft adapted the CSI 16-Division system for categorizing symbols, which it calls its Object Library; ArchiCAD objects have the extension GSM. Look for them in folder ArchiCAD\ArchiCAD Library\ Object Library.

Revit provides its symbol library in both Imperial and metric formats; the symbols have the extension RFA. Look for them in the folder called Revit\Data\ Imperial\ Library.

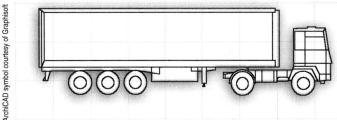

A few vendors provide such a large library that a second CD-ROM is required. TurboCAD, for example, has nearly 12,000 2D and 3D symbols taking up 366MB in British, Internation Standards Organization (ISO), and United States formats. Symbols are stored on the CD-ROM in files with the TCW extension.

A/E/C CADD Symbology Libraries

Your clients may require that you make use of their symbol library. This ensures uniform drawing standards. One example is the CADD/GIS Technology Center, which provides symbols in AutoCAD and MicroStation format at tsc.wes.army.mil/products/standards/aec/Sym-index.asp

In addition to symbols, the site has linetype and hatch pattern files.

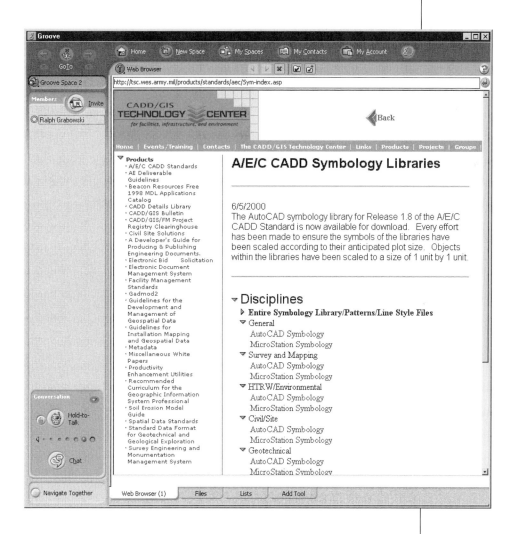

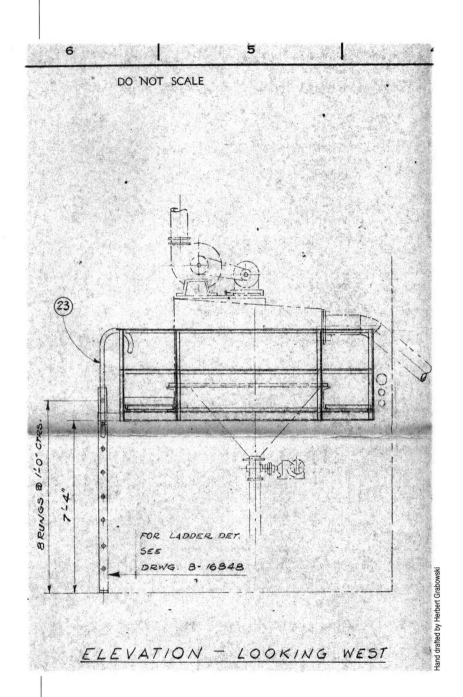

Layer Names and Conventions

3

All CAD packages give you the freedom to create many layers in the drawing. This leaves the new CAD user in a quandary: What do I do with layers?

Even though color is the subject of the next chapter, layers and colors are so closely intertwined that you'll notice the occasional reference to color in this chapter.

WHAT ARE LAYERS?

If you are familiar with overlay drafting, then it is easier to understand layers in CAD. Overlay drafting lets you combine several Mylar drawings to create a master blueprint. For example, to produce blueprints for the plumbing contractor, you combine the Mylar sheets of the site plan, the structural plan, and the plumbing plan. The plumber is not interested in seeing the furniture placement or the landscaping plan.

In CAD, layers let you have all elements of a project in a single drawing: site, structural, plumbing, furniture, and landscaping. You place each on a separate layer and then turn layers on and off to create plotted drawings as required. Turn off the furniture and landscaping layers to plot the drawing for the plumbing crew.

Layers in CAD

Most CAD systems allow you to create an unlimited number of layers in each drawing. Others are more limiting. For most of

CHAPTER SUMMARY

This chapter describes approaches to implementing a layer-naming scheme for your CAD drawings. The chapter includes descriptions of several industry standards:
- AIA CAD Layer Guidelines
- CSI MasterFormat
- CSI UniFormat
- U.S. Coast Guard
- CalTrans Drawing Levels
- ISO Layers

its life, MicroStation limited drawings to 63 layers (MicroStation 8 will have an unlimited number of levels).

The earliest version of AutoSketch supported a mere ten layers.

Most CAD packages refer to layers by name; a few are limited to assigning a number, typically ranging from 0 to 255. Layer names might be limited to six or eight characters, but it is more common now to allow names as long as 255 characters, as in AutoCAD.

An increasing number of CAD packages have the ability to create a layer *group* (sometimes called "layer sets"). This allows you to toggle or freeze the display of more than one layer at a time. Layer groups also allow you to work with logical sets of layers, rather than be limited to viewing their names in alphabetical order. For example, you could work with all layers related to landscape design. This becomes important in drawings that externally reference numerous other drawings, which results in a drawing containing hundreds upon hundreds of layers!

A few CAD packages have the ability to import and export layers in and out of their drawings.

Another handy feature is the ability to print out the layer tables. If the CAD package does not directly support these two features, sometimes third-party applications are available to do the job.

To learn how to create and modify layers in specific CAD packages, refer to the software's documentation.

A CAD PACKAGE WITHOUT LAYERS

Revit doesn't allow you to create or set layers; it doesn't even have the concept of "layers." Instead, as you place objects, Revit automatically places them in categories and subcategories.

Before exporting the drawing in DWG or DXF format, you can map Revit categories to layer names. By default, Revit maps its categories to a subset of the AIA's layer system via the ExportLayers.Txt file in the \Revit\Program folder.

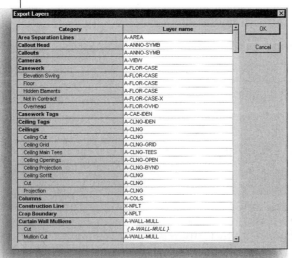

HOW TO NAME A LAYER

Most CAD systems are unhelpful when it comes time to decide on layer names. Like a doctor who delivers your baby, but doesn't name it, CAD vendors are happy to deliver the software, but leave layer naming up to the customer.

AutoCAD, for example, allows an unlimited number of layers, yet it defines a single layer named "0" in its default template drawing — most unhelpful for the neophyte CAD user! AutoCAD includes many template drawings based on standards from **ANSI** (American), **ISO** (international), **DIN** (German), and **JIS** (Japanese) standards organizations. These, however, include just a few rudimentary layer names, such as Viewport, (drawing) Frame, and Titlebock.

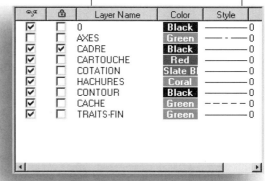

TurboCAD's display of layers in a dialog box.

TurboCAD is almost as sparse in naming layers, but does include a pair of template drawings — Drawing Board.tct and Drawing Board Metric.tct — that includes a rudimentary layer naming system: Dimensions, Text, Hatches, Points, Lines/Arcs/Curve, $Construction, and Border. Better yet, macros stored in the template drawings cause objects to be drawn on the appropriate layer, automatically. ($Construction is a non-plotting layer for drawing construction lines.)

MicroStation is more helpful to the user by including the AIA's CAD layer guidelines as an optional layering convention. As this book was being written, a press release from Bentley Systems indicated they would be adding the CSI layer standard to the next release of MicroStation.

Cadvance makes 255 layers in new drawings, named "Layer 1" through "Layer 255." Cadvance v99.5 includes a single template file called **Lib_title.vwf**, which contains layers based on the AIA's *CAD Layer Guidelines* for architecture, such as A-WALL-FULL. Additional discipline-specific layers can be found in the following drawing files:

- **Lib-plum.vwf**   Plumbing layers, such as P-ACID-PIPE
- **Lib-lite.vwf**   Electrical layers, such as E-LITE-SPCL
- **Lib-grid.vwf**   Structural layers, such as S-FNDN-RBAR

CadKey drawings have 255 layers numbered 1 through 255. Although the layers cannot be renamed, a description can be attached to each layer number.

Most CAD systems, however, present you with a blank sheet. For this reason, I present five strategies for naming layers:

Strategy #0: Do Nothing

When you use CAD infrequently, don't bother setting up a color or layer naming system. Draw everything in black and draw them on layer 0. For simple drawings, it's a waste of time to toggle colors and switch layers.

Strategy #1: The Simple Plan

When you use CAD in-house for simple drawings, create a simple color and layer system that makes sense for the kind of drawings you produce. At the very least, every drawing should have layers for text, hatching, and dimensions. You want text, hatch, and dimension entities on separate layers to allow you to suppress the display of extraneous detail if you decide to use the drawing as an external reference. Keeping the information on separate layers also gives you more control over the appearance of plots.

Call the layers "Text", "Hatch", and "Dim" and color them white. If your CAD package doesn't allows named layers, use layer numbers 1, 2, and 3 — it's that simple. Don't use layer 0 since many CAD packages use that layer for special purposes.

Strategy #2: The Plotter Plan

You may find that drawing and plotting is easier if the layers match your plotter's "pens." (Although modern plotters don't use pens to draw lines, it's convenient to say "pen width" instead of "line width produced by the plotter.") Here is a layer system that matches layers with the most common pen widths:

Pencil Drafting	Object Color	Layer Name	Pen Width
Soft: 7B to 2B	White	1	Bold: 0.70 mm or 0.03"
Medium: B, HB, F, H	Yellow	2	Medium: 0.50 mm or 0.02"
Hard: 4H to 9H	Green	3	Extra Fine: 0.25mm or 0.01"
Medium: 2H to 3H	Any other color	4	Fine: 0.35 mm or 0.015

Some CAD packages (such as AutoCAD and MicroStation) show line weights on the screen and let you match line weights to pen numbers.

> **TIP LAYERS VERSUS LEVELS**
> While most CAD packages use the term "layer," MicroStation uses the term "level."

MOST COMMON STANDARDS ORGANIZATIONS

ANSI
American National Standards Institute
www.ansi.org

ISO
International Organization for Standardization
www.iso.ch

DIN
Deutche Industrie Norm (Germany)
www.din.de

JIS
Japanese Institute of Standards
www.hike.te.chiba-u.ac.jp/ikeda/JIS/

Strategy #3: The Four-Step Plan

Depending on the type of drawing your firm produces, you may want to create layers in addition to the basic three. Here are the steps to designing an in-house layer-naming system:

Some layer-naming systems are quite straightforward.

Step 1. Gather together a number of drawings and scribble down logical elements you find in all of them. I've already mentioned text, hatching, and dimensions. Other elements include: electrical, plumbing, HVAC, LAN, NC tool paths, plasma cutter, sprinkler locations, conifers, and shrubs.

Step 2. Create a layer name that describes each element, such as "Plumbing", "Electrical", and "HVAC." Now shrink each potential layer name to fit your CAD package's naming limitation. For example, if your CAD package limits names to eight characters, "Plumbing" fits the limit but "Electrical" must to be chopped back to "Electric". (If your CAD package uses numbers, then create a cross-reference that lists the layer number and its meaning. Post the list at each CAD station.)

Step 3. Now think about subcategories. If your firm does renovation work, you need more than just one layer for plumbing. Drawings must show three kinds of plumbing — existing to remain, existing to be removed, and new — and each should be on its own layer. In a CAD package without the short name limitation, you can afford to be descriptive: "Plumbing-remain," "Plumbing-remove," and "Plumbing-new."

To fit an eight-character limit, this step involves a second round of layer name shrinking: "PbRemain", "PbRemove", and "PbNew".

If your CAD package uses numbers for layer names, consider logical groupings. For example, all existing elements could be drawn on the 10-group of layers, construction drawn on the 20-group, and all maintenance on the 30-layers. Within each group of ten layers, apply the individual disciplines, such as x1 for plumbing. You end up with layer 11 as the existing plumbing, layer 21 for the plumbing to be constructed, and layer 31 as the layer for plumbing maintenance.

PRINTING AUTOCAD LAYERS

AutoCAD cannot print a reference of layers. DotSoft has a product that presents all the info found in AutoCAD's layer dialog box in HTML format. LayerHtm is freeware available from www.dotsoft.com

Layer Names and Conventions

Step 4. Finally, create a template (or "seed" or "prototype") drawing with the layers and colors you've defined. Each time a CAD operator begins a new drawing, the layers and colors are predefined. See Chapter 7 for more information on creating a template drawing.

The following table is helpful in developing a simple in-house layer naming convention.

LAYER NAMES FOR IN-HOUSE DRAWINGS

Architectural	GIS
Dim	Elec
Grid	Esca
Misc	Cont
Text	Lege
Symbol	Text
Ceiling	Title
Floor	Vent

Strategy #4: Do What Your Client Says

To be compatible with your clients, your firm may be forced to adopt their layering standards as discussed in the following section.

Strategy #5: Copy What Works, Make Minor Modifications

If your client doesn't have a CADD standard, you'll be doing them a favor if you clone the CADD standard of a similar major client. For instance if you're working on a canal, clone the U. S. Army Corp of Engineers. If you're designing a road in Upper Volta, clone your local State DOT (Department of Transportation) standard. If you stick to a widely known standard, you'll have fewer difficulties coordinating with government agencies, and other AEC firms. You'll probably be able to use existing symbol libraries, drawings, and mapping. And you'll be less likely to discover serious errors in the logic of the CADD standard. Don't be a slave to the existing standard; make minor changes if it would benefit your client.

Some standards are listed in the following sections.

THE STRUGGLE TO CREATE LAYER STANDARDS

Starting in the late 1980s, a number of organizations began the task of creating a layering standard, as well as the related colors and linetypes (discussed in the chapters following). The overriding concern behind this effort was drawing exchange. To better understand the problem layers can create, consider the following scenarios.

Several engineering firms are working together on a large project. Although each is using MicroStation, each uses its own layer system. When it comes time for one firm to read in the drawings of another firm, a problem arises: the layer names conflict with each other. What one firm calls the "02060" layer, another firm calls the "Building Demolition" layer, and the third firms calls the "A-WALL-DEMO" layer.

Clearly, if the firms had agreed upon a layer-naming standard, then they would be able to easily pass their drawings to each other. After more than a decade of effort, however, it has become clear that developing a single standard for naming layers in CAD drawings is an impossible task. Some of the barriers to creating one all-encompassing standard are:

Limitations of CAD Systems

CAD software itself may pose a limit on the number and type of layer names. At one extreme, an AutoCAD drawing is capable of holding an unlimited number of layers, each with a name up to 255 characters long. Other CAD systems limit the number of layers (such as 10, 63, or 256 layers) per drawing, or the length of the name (six or eight or twelve characters), while a few CAD systems use hardcoded numbers — 0 through 255. Any layer-naming standard that caters to both extremes would be limited by the lowest-common denominator, which would fail to serve the users of more flexible CAD systems.

Conflict of the Disciplines

A layer-naming system that meets the need of one discipline may not meet the need of another discipline. Names suitable for layers in a mechanical drawing are unsuitable for an architectural drawing. Both would be different from the needs of a dress designer. While some CAD systems could hold all possible layer names, less capable software would need to split the layer stan-

Icons used by AutoCAD to display the status of layers. From top to bottom:
On/off
Thawed/Frozen
Layout
Unlocked/locked

dard into disciplines, or create a system that handles 63 or 256 layers at a time. MicroStation does that through the use of reference files.

Human Quirks

Debate in the CAD journals has shown that users have different preferences in approaches to layer standards. Some prefer alphanumeric-based layer names; others prefer numerical layer names. Others prefer a complex system that lets you separate out nearly any unique aspect of the drawing; a few prefer a simple system that simply matches half-a-dozen layers with plotter pen colors. Some prefer layers subdivided by subcontractor; others prefer layers subdivided by building material.

With these barriers in mind, organizations in several countries have forged ahead and created layer-name standards. They overcame the barriers previously listed by: (1) making them discipline-specific; and (2) making allowances for shortened or numbered layer names. As for human quirks, advances in graphical user interfaces, together with macros, have obviated the need for CAD operators to type layer names. Indeed, it is possible to customize a CAD system so that objects are placed automatically on the required layer.

With this background in mind, here are excerpts from four layering systems. I give examples from four kinds of systems:

- A word-based system from the AIA
- A number-based system from the CSI
- A mixed letter-number system from the CSI
- An adaptation of the CSI system by the U. S. Coast Guard
- A homegrown system from CalTrans
- An adaption of the AIA system by the ISO

There simply isn't room in this book to document each standard completely; each contains so many layer names that it comes in its own book (or on a CD-ROM)! The AIA system, for example, contains nearly 1,500 layer names.

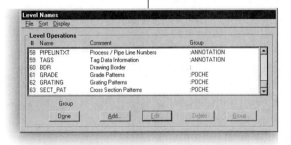

Although MicroStation/J was limited to 63 levels (layers), it provided better control over those layers than most other CAD systems. Notice how levels can be assigned by Name, Comment, and Group. Toggling level groups is a fast way to view specific parts of a drawing.

AMERICAN INSTITUTE OF ARCHITECTS

In the mid-80s, the AIA set up a task force on CAD layer guidelines. Their job was to create a layer naming system for building design, construction, and facility management. The first version came out in 1988; it has since been revised a couple of times. The system is not intended for other kinds of CAD drawings, such as mapping, highway design, printed circuit design, process plant engineering, or aircraft design.

The naming system uses multiple fields separated by hyphens called "delimiters." The fields divide the layer name into logical sections. For example:

E-LGHT-EMER-NEWW
E-LGHT-SWCH-DEMO

These two layer names for electrical emergency lighting to be added and light switches to be removed.

Notice how the delimiters separate the layer name into categories:

Category	Example	Meaning
Major category:	E	Electrical
Group:	LGHT	LiGHTing
Subgroup:	EMER	EMERgency
Modifiers:	NEWW	NEW Work

The task force arrived at a naming system to accommodate CAD packages with long layer names (thirty-one characters or more), short layer names (typically eight characters), and those limited to numbered layers:

- **Long-form** names have four or five fields, such as E-LGHT-EMER-NEWW
- **Short-form** names consists of the first two fields, such as E-LGHT or EXLGHT
- **Numbered** layers assign numbers to specific layers, such as 130 for the emergency lighting layer.

> **TIP WILDCARDING LAYER NAMES**
>
> When we named the three plumbing layers, we began each with the same letters, such as "Pb..." or "Plumbing...". That was for a good reason. All CAD packages (except MicroStation) let you turn layers off and on individually or by groups by using wildcards:
>
> ***** (*asterisk*) matches a group of characters
>
> **?** (*question*) matches a single character
>
> You probably recognize these characters as the same wildcards used by DOS and Unix. You quickly turn off (or turn on) all the plumbing layers by referring to them as "Pb*" or "Plumbing*".
>
> The numbered layers are handled slightly differently. Since the *second* digit defines a similar group, use "?1" to turn off or on all plumbing layers.
>
> TurboCAD has a handy feature that lets you specify layer name prefixes, which saves a bit of time creating many new layers that start with the same several characters.

Major Categories

The AIA's system has eight major categories, which represent specific disciplines in the construction of a building:

Major Category	Abbreviation
Architecture	A
Structural	S
Mechanical	M
Plumbing	P
Fire Protection	F
Electrical	E
Civil	C
Landscape	L

In addition to these eight, there is a ninth category specific to the drafting process:

Drafting Process	Abbreviation
Break line	BRAK
Center lines	CNTR
Cross hatching and poche	PATT
Dimensions	DIMS
Drawing revision number 1	REV1
Elevation	ELEV
Floors	FLRn *
General notes and specs	TEXT
Hatching	HTCH
Identification tags	IDEN
Nonplot info (construction lines)	NPLT
Notes, call-outs and key notes	NOTE
Options	OPTn *
Phases	PHSn *
Plotting targets and windows	PLOT
Section marks	SECT
Solid shading	SHAD
Symbols, bubbles, targets	SYMB
Tags	TAGG
Text labels	LABL
Title block sheet name/number	TTLB

* n is replaced by a number, such as FLR1 for the first floor.

Groups and Subgroups

The AIA's *CAD Layer Guidelines* contain too many groups and subgroups to list here; they would take up sixty pages of this book. To give you a taste, here are some groups in from the architectural discipline:

Description	AIA Layer Name
Full height stair shaft walls	A-WALL-FULL
Partial height walls	A-WALL-PRHT
Movable partitions	A-WALL-MOVE
Door and window headers	A-WALL-HEAD
Door and window jambs	A-WALL-JAMB
Wall insulation hatch fills	A-WALL-PATT
Wall surfaces—3D views	A-WALL-ELEV

WALL is the name of a group, and FULL is an example of a subgroup.

Modifiers

Generally, everything in a construction project is either built new, existing (but to be demolished), or existing (and remains). To specify these, you add the following modifiers to the end of the layer name:

Meaning	Modifier
New work (construction)	NEWW
To be demolished	DEMO
Existing	EXST

The complete *CAD Layer Guidelines* are available for purchase from the AIA.

RESOURCE

To obtain a copy of the AIA's *CAD Layer Guidelines* book, search www.e-architect.com.

Additional information is available at www.mcs.net/~djec/rclg/aia_clg1.htm

CSI MASTERFORMAT

The Construction Specifications Institute (CSI) is the driving force behind creating organization out of the chaos of the construction industry. The MasterFormat system is almost universally employed to organize information about construction materials; you'll find it used at many construction-oriented Web sites and catalogs.

Sometimes called the "16-Division" system, the CSI layer naming system uses six-digit numbers to separate a CAD drawing into layers based on construction workflow. Each division represents a group of construction specifications, such as site work (02), concrete (03), and finishes (09).

The original 16 divisions were created in 1988, and are called "MasterFormat." Although it was not designed for layers, nevertheless some companies find it convenient to use as a layering system.

In 1995, a seventeenth division was added, Division 0 for introductory material, such as bidding and contracting requirements.

MasterFormat does not support facilities, spaces, systems, or assemblies. The full list of divisions is as follows:

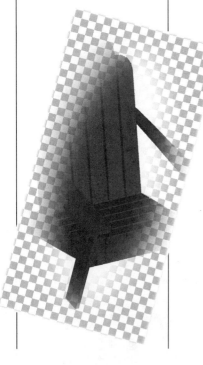

Division	Meaning
0	Introduction and Requirements
1	General
2	Site
3	Concrete
4	Masonry
5	Metals
6	Wood and Plastics
7	Thermal and Moisture
8	Doors and Windows
9	Finishes
10	Specialties
11	Equipment
12	Furnishings
13	Special Construction
14	Conveying Systems
15	Mechanical
16	Electrical

For example, the following layer number represents existing emergency fixtures to remain on floor 3:

16531-3

The first two digits, called the "Basic Divisions," represent 16 CSI divisions, numbered from 00 to 16. Here, division 16 represents electrical.

The second pair of digits, referred to as the "narrow scope numbering," represent the details within each division, such as roads (02*52*), gypsum board (09*25*), and furniture (12*62*). Emergency fixtures are 1653.

The fifth digit defines the status of the construction material, as follows:

Status	Meaning
xxxx1	Existing to remain
xxxx2	Existing to be removed
xxxx3	New
xxxx4	Text
xxxx5	Dimensions
xxxx6	Hatching
xxxx7 — xxxx9	User definable

Thus, existing emergency fixtures to remain are placed on layer 1653*1*.

The final pair of characters are optional and designate the floor number of a multistory project. The dash (-) is a delimiter, while the 3 represents the third floor.

For simple drawings, the system encourages you to use just the first two digits as layer names.

The advantage to the 16-division layer system is that it matches the CSI construction specification book. The logical structure makes it easy to toggle layer visibility with wildcard characters. The disadvantage is that the numbers are meaningless to a CAD operator unfamiliar with the CSI spec.

Like the AIA's layer specification, the complete CSI spec is too long to reproduce in this book. Appendix C lists the light version provided by the CSI, which lists just the "level two" numbers (there are four levels in total) — over 200 layer names!

RESOURCE
To obtain a copy of the CSI's *MasterFormat* book, search www.csinet.org.

CSI UNIFORMAT

In addition to MasterFormat, CSI developed another specification system called UniFormat, first developed in 1992. The idea behind this system is to define the physical parts of a facility, which the CSI calls "systems and assemblies."

Instead of describing the building materials (as does MasterFormat), UniFormat describes the function of the building parts. UniFormat is intended to complement MasterFormat, as the following example illustrates. UniFormat uses a series of letters and numbers to specify systems and the MasterFormat divisions: The following table shows how to interpret the UniFormat string "*A*1010.03300X":

CSI Level	Field	Value	Meaning
Level 1	**A**1010.03300X	A	Substructure
Level 2	A**10**10.03300X	10	Foundations
Level 3	A10**10**.03300X	10	Standard Foundations
Level 4	A1010.**03300**X	03300	MasterFormat
Level 5	A1010.03300**X**	X	User defined

The standard assigns numbers and letters to levels 1, 2, and 3. Level 4 can be employed for the MasterFormat number, while level 5 is available for the user. In the previous example, 03300 refers to the MasterFormat number for Cast-In-Place Concrete. The user-defined level could be used to specify floor or building numbers.

Level 1	Meaning
A	Substructure
B	Shell
C	Interiors
D	Services
E	Equipment and Finishings
F	Special Construction and Demolition
G	Building Sitework
Z	General

RESOURCE
To obtain a copy of the CSI's *UniFormat* book, search www.csinet.org.

Like the MasterFormat specification, the complete UniFormat spec is too long to reproduce in this book. Appendix C lists the light version provided by the CSI.

US COAST GUARD

The CSI's MasterFormat system has been adapted as a layering system by the U. S. Coast Guard. They added two more divisions: Division 00 is dedicated to contour elevation lines, while Division 20 is for reference.

Layer	Division
00000	Contour Lines
01000	Field Engineering
02000	Sitework
03000	Concrete
04000	Masonry
05000	Metals
06000	Wood and Plastics
07000	Thermal and Moisture Protection
08000	Doors and Windows
09000	Finishes
10000	Specialties
11000	Equipment
12000	Furniture
13000	Special Construction
14000	Conveying Systems
15000	Mechanical
16000	Electrical
20000	Reference

To indicate the status of a layer, three codes can be added to the end of the layer name:

Status Codes	Type Of Work
00	New Design
01	Existing Design
02	Demolition

For example, when you need to add a foundation to an existing structure, draw the foundation on the 01 layer. If you need to demolish the existing structure, show this on the 02 layer, as follows:

Primary Layer:	03650	FOUNDATION
Create Layer:	0365**01**	EXISTING FOUNDATION
Demolish Layer:	0365**02**	FOUNDATION TO BE DEMOLISHED

Layer Names and Conventions

The Coast Guard added a category not found in the CSI specification for contour lines — Division 0. Contour lines are placed on one of four layers:

Layer	Elevation
00001	Elevations divisible by 1.0
00002	Elevations divisible by 2.0
00005	Elevations divisible by 5.0
00010	Elevations divisible by 10.0

Here are samples of layers for this contouring scheme:

Layer	Sample Elevation
00001	533
00002	434
00005	625
00010	740

In addition, markers are placed on layer 00003.

The Coast Guard uses the following Division 20 layers for placing reference lines:

Division 20	Reference
20100	Centerline (Center)
20102	Centerline (Center3)
20105	Col. Balloon, Col. Centerline (Center2)
20200	Phantom Line (Phantom)
20202	Phantom Line (Phantom3)
20205	Match Line (Phantom2)
20300	Broken Line (Hidden)
20302	Broken Line (Hidden3)
20305	Broken Line (Dashed)
20400	Solid Line, Continuous
20500	Leaders, Leader Text
20600	Dimensions, Notes, Text
20700	Miscellaneous
20800	Revision Cloud, Tag
20900	Crosshatch

CALTRANS DRAWING DATA LEVELS

The California Department of Transportation *CADD Users Manual* is an example of the guideline consultants must use to ensure uniform procedures for creating roadway drawings.

CalTrans and its consultants create drawings by merging master drawings in various combinations. Since the drawings are based on MicroStation, the level (a.k.a layer) names are numbered between 1 and 63.

Layers are grouped as follows:

Levels	Meaning
1 - 8, 12	Basic topographic map data
9, 10	Sheet formats
13 - 30	Basic construction details
31 - 35	Right-of-way data
36 - 59	Data specific to type of plan sheet
60	Non-geographical drawing data
61	Final plan revisions
62	As-built changes

Levels 11 and 63 are not used.

There is no advantage or disadvantage to the CalTrans layer system; it is simply the system your firm must use if producing drawings for CalTrans. CalTrans doesn't care what COGO (coordinate geometry) package you use, as long as you can produce the DGN files; most consultants use InRoads.

RESOURCE

For more information, go to svhqsgi4.dot.ca.gov:80/hq/esc/Engineering_Technology/DevelopmentBranch/webpage.htm

ISO LAYER STANDARD

The International Organization of Standards (ISO, headquartered in Switzerland) created in 1988 a layer standard named ISO 13567-2 based on the AIA's CAD *Layer Guidelines*. As an example, the layer that specifies the text for identifying a room's name is:

In AIA format:	A-FLOR-IDEN
In ISO format:	A-FLORIDM-NB10131FRC

The dash (-) is used as a placeholder. ISO layer names consist of two parts: ten required characters, followed by ten optional:

Mandatory	A-FLORIDM-
Optional	NB10131FRC

Mandatory Part

The mandatory part of the ISO layer name consists of three parts:

Agent responsible	A-
Element	FLORID
Presentation	M-

Agent Responsible: The first two characters are based on the AIA's discipline code. For example, A- for architecture.

The second character is optional, and indicates a specialization. For example, AG for architectural graphics, or A1 to specify the architect number.

Element: The middle six characters can be defined by each country. One system uses a condensation of the AIA's major and minor groups. All four characters of the major group are used, along with the first two characters of the minor group. For example, FLORID (shorted from the AIA's FLOR-IDEN).

The following common modifiers are defined for use in the minor group field:

Minor Group	Meaning
DE	to be DEmolished
EX	EXisting to remain
ID	IDentification
NW	New Work
PA	PAttern

Another system, based on the Swedish building element classification system BSAB96, looks like this:

Swedish Layer Group	Element
Landscape, general	100
Building, general	300
Loadbearing walls	331
Roof	340
Exterior walls	350
Openings in exterior walls (doors, windows, etc.)	355
Interior (non loadbearing) walls	363
Suspended ceilings	364
Stairs	366
Interior surfaces	370
Spaces	900

Presentation: The final two mandatory characters indicate whether the layer is part of specific elements. For example, the layer for general notes and general remarks is:

A - - - - - NO**P** -

The P indicates the layer is part of the *page* (a.k.a. the layout, paper space, or sheet). When an M appears instead, the layer is part of the model.

ISO specifies the mandatory characters for the presentation portion of the layer name. The first character codes are listed below; the second character is optional, and can be a number.

ISO Presentation Layer	Code
Whole model and drawing page (two hyphens)	--
Model	M
Element graphics	E
Annotation	A
Text	T
Hatching	H
Dimensions	D
Section/detail marks	J
Revision marks	K
Grid	G
Graphic	Y
Dimension	Z

MODEL VERSUS SHEET

Model layers are usually scaled, while sheet layers are full size (1:1).

ISO Presentation Layer (continued)	Code
User	U
Redlines	R
Construction lines	C
Page/paper	P
Border	B
Border lines (Frame)	F
Other graphics	O
Text	V
Title	W
Notes	N
Tabular Information	I
Legends	L
Schedules	S
Tables (Query)	Q

Optional Part

The optional part of the ISO layer name consists of seven parts:

ISO Optional Layer Name Parts	Length
Status	One character
Sector	Four characters
Phase	One character
Projection	One character
Scale	One character
Work Package	Two characters
User Defined	Any number of characters

Status: The first optional character indicates the status of the layer:

ISO Status Designator	Meaning
D	Demolition
E	Existing to remain
F	Future work
M	Items to be moved
N	New work
X	Not in contract
R	Relocated items
T	Temporary work

RESOURCES

In addition the AIA, CSI, USCG, and CalTrans, you may want to check out standards at the following Web sites:

US Army Corp of Engineers Standards
(Seattle District)
www.nws.usace.army.mil/cadd/cadd.htm

City of San Diego
(MicroStation format)
www.sannet.gov/engineering-cip/drawings.shtml

Sector: The next four characters indicate the floor number or other building sector.
Phase: The sixth digit indicates any sort of phase of the project, and ranges from 1 through 9.
Projection: The seventh digit indicates the projection, such as plan, section, and elevation.
Scale: The eighth letter indicates the drawing scale. Some examples include:

ISO Scale Designator	Meaning
D	1:20
E	1:50
F	1:100
G	1:200

Work Package: The final two characters indicate the work package, such as materials or products.
User Defined: Codes can be added to the end of the layer name, consisting of any number of letters and numbers

The complete 13567 layer standard is available for purchase from the ISO.

RESOURCES
For more information about the ISO layer standard, check these Web sites:

ISO
www.iso.ch

ISO 13567
www.proteus.ie/iso.html

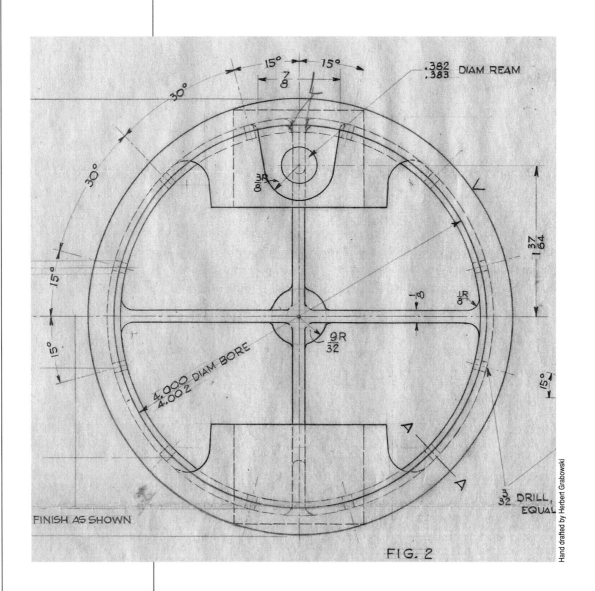

FIG. 2

Assigning Colors

4

All CAD packages give you the freedom to use many colors in a drawing. Whereas CAD systems of yesteryear limited you to 255 or even just fifteen colors, today's CAD systems support the 16.7 million colors. That leaves new CAD users in a quandary: What color coding system? Why use colors at all?

COLORS

Why use color in a CAD drawing? Color was rarely used in manual drafting, so you may question the thought of using color in an electronic drawing. Feel free to do so. Since the output from a CAD system is, in most cases, a black-white plot, you may not want to use color.

The Case for Color

CAD programs use color for two purposes: (1) operator cues, and (2) controlling the plotter.

When a drawing contains colored elements, the colors provide cues to the operator, making them more efficient. It's easier to pick out a blue water pipe from a sea of red structural members than when pipes and steel beams are both drawn in black.

While the structural engineering firm decided against a color display, they still need to use color in drawings — even if the colors are not displayed or plotted. That's because all CAD systems use color to control the plotter's pens.

CHAPTER SUMMARY
This chapter describes how to use colors in CAD drawings. It includes approaches to creating a standardized color scheme, and describes several industry color standards.

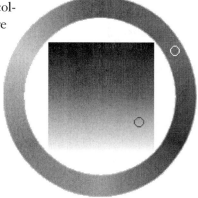

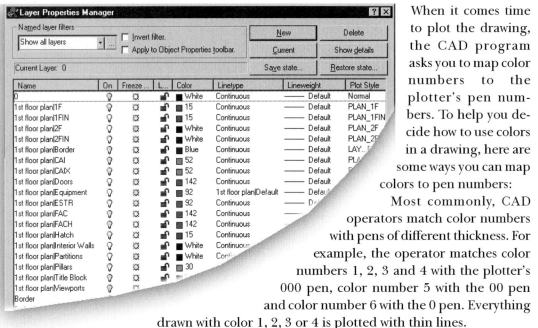

Assigning colors to layers via a dialog box.

When it comes time to plot the drawing, the CAD program asks you to map color numbers to the plotter's pen numbers. To help you decide how to use colors in a drawing, here are some ways you can map colors to pen numbers:

Most commonly, CAD operators match color numbers with pens of different thickness. For example, the operator matches color numbers 1, 2, 3 and 4 with the plotter's 000 pen, color number 5 with the 00 pen and color number 6 with the 0 pen. Everything drawn with color 1, 2, 3 or 4 is plotted with thin lines.

When color output is needed, the CAD operator matches the color number with the plotter's pen number. For example, color number 1 is matched with the red pen. Everything drawn with color 1 is plotted in red.

Monochrome plotters work with shapes of gray, rather than pen numbers.

Naturally, you can map all colors to pen #1 for draft plots or for output to a monochrome laser printer.

The Case for Monochrome

One structural engineering firm did just that. When it came time to upgrade their CAD hardware, the firm's principals got the CAD operators involved. Since the operators are the ones who use the equipment, it made sense to let them have a say in the selection of the graphics display.

The operators decided on a very-high resolution monochrome display system for two reasons: (1) the final output would always be monochrome, and (2) they would be more productive since the extra-high resolution meant less time-consuming zooms and pans. The money saved on losing color allowed the firm to spend that money on higher resolution and bigger screens.

> **TIP LIGHT GRAY SCREENING**
>
> Some plotting software allows users to apply a "screen" to objects based on layer/level. Screened objects are plotted in a light gray shade, which makes them less conspicuous. Typically existing features, such as contour lines, are screened, while new features are unscreened.

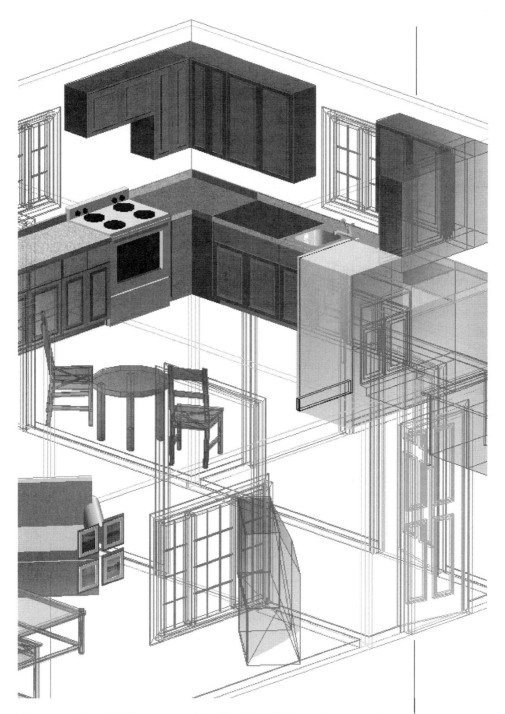

Consumer-oriented CAD systems, such as FloorPlan 3D shown here, preassign colors and textures to predrawn objects. Other CAD systems use color numbers to assign textures to surfaces.

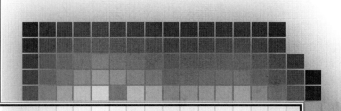

TIP COLOR NUMBERING SYSTEMS — WHAT THEY MEAN

RGB: Red-green-blue is the system used by Windows. Each of the three primary colors — red, green, blue — range in intensity from 0 (black) to 255 (full color). This system allows Windows-based software to specify 16.7 million colors (256 x 256 x 256).

HSL: Hue-saturation-luminance is an alternative color specification system, where each parameters also ranges from 0 to 255. Hue starts with red (0) and goes through other colors: yellow (43), green (85), blue (170), and back to red (255). Saturation is the intensity of the color, from 0 (gray) to 255 (full color). Luminance is the brightness of the color, ranging from black (0) to white (255).

EGA: Named after IBM's enhanced graphics adapter, and is used by many CAD systems.

HTML: Hypertext markup language, and is the color specification system used by Web sites. It is identical to the RGB system, but is in hexadecimal (base 16) notation; the # prefix is a reminder that these values are in hex: 0,1,2 ... 8,9,A,B,C,D,E,F (the letters A through F represent the numbers 10 through 15). In hex notation, #F is 15, #FF=255, and #80=128.

DGN: The color numbering system used by Bentley Systems' MicroStation.

ACI: Autodesk's AutoCAD Color Index, and is used by AutoCAD and AutoSketch.

These are not the only color systems. Others include:

CYMK: Specifies the amount of cyan, yellow, magenta, and black in the range of 0 to 255. This is the system used by color printers.

Crayola: Yes, one CAD package even used the 128 color names found in the Crayola crayon package.

HOW CAD WORKS WITH COLORS

CAD software doesn't work with color names but with color *numbers*. The software matches the number to a color. (Sometimes, the CAD software lets you specify a color name as a pseudonym for the number.)

There are different ways of matching the number with the color displayed on the screen. The following table shows the color names for the first fifteen color numbers for several color systems:

COMPUTER COLOR NUMBERING SYSTEMS

Name	RGB	HSL	EGA	HTML	DGN	ACI
Black	0,0,0	0,0,0	0	#000000	255	0
Blue	0,0,255	170,255,128	1	#0000FF	1	5
Green	0,255,0	85,255,128	2	#00FF00	2	3
Cyan	0,255,255	128,255,128	3	#00FFFF	7	4
Red	255,0,0	0,255,128	4	#FF0000	3	1
Magenta (pink)	255,0,255	213,255,128	5	#FF00FF	5	6
Yellow	255,255,0	43,255,128	6	#FFFF00	4	2
White	255,255,255	0,0,255	7	#FFFFFF	0	7
Gray	128,128,128	0,0,128	8	#808080	9	8
Light Blue	128,128,255	170,255,192	9	#8080FF		13
Light Green	128,255,128	85,255,192	10	#80FF80		11
Light Cyan	128,255,255	28,255,192	11	#80FFFF		12
Light Red	255,128,128	0,255,128	12	#FF8080		9
Light Magenta	255,128,255	213,255,192	13	#FF80EE		14
Light Yellow	255,255,128	42,255, 92	14	#FFFF80		10
Light Gray	192,192,192	0,0,192	15	#C0C0C0		15

AutoCAD matches a number to each of 255 colors. The first fifteen are listed under ACI (AutoCAD color index) in the preceding table.

MicroStation is similar, but uses a different set of numbers, as listed under DGN in the table. In addition, MicroStation allows you to change the color associated with each number.

TurboCAD specifies colors via the RGB and HSL systems, as shown in the illustration at right.

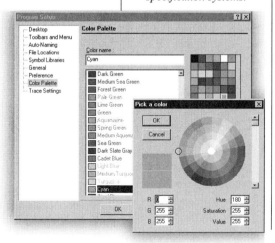

TurboCAD allows the user to select colors from a palette of 16.7 million via the RGB and HSL color specification systems.

Assigning Colors

FOR AUTOCAD USERS

When you specify 0 for **Use Assigned Pen Number**, AutoCAD determines the plotter pen of the closest color to the object you are plotting using the information you provided under **Physical Pen Characteristics** in the **Plotter Configuration Editor**.

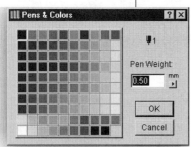

ArchiCAD preassigns a pen number and width to every one of its ninetynine colors.

WHERE TO ASSIGN COLORS

In general, there are two ways you can assign colors in a CAD drawing: (1) by layer; and (2) by object.

The correct method is to assign color by layer. The reasoning is that you usually draft common elements on a single layer, which logically have the same color. In addition, when you change the color of a layer, the CAD system then automatically changes the color of all objects on that layer.

AutoCAD uses a special color name: color **Bylayer** means the object takes on the color assigned by the layer. The illustration shows how colors are assigned to layers in AutoCAD's **Layer Properties** dialog box.

The incorrect method is to override the layer color and specify the color of individual objects in the drawing. While this seems more intuitive, it gets to be a real mess when you need to change colors later.

Matching Colors to Pens

In the early days of CAD, you would match color numbers to pen numbers at plot time. This indirect system allowed any physical pen to be used by the CAD system. For example, when CAD color number 1 was assigned (which might be red) to plotter pen #1, the plotter used whatever pen was in holder #1, whether red or black, thin or wide.

Some systems allow you to assign one CAD color to multiple plotter pens. For example, you could match CAD color 1 to physical pens 1 and 2. During a long plot, when the first pen runs out of ink, the plotter switches to pen #2.

ArchiCAD, for example, assigns each color to a pen number and a pen width via the **Options | Pens & Colors** command. All colors have been preassigned pen widths, but these can be edited by the user.

Some of today's CAD systems go far beyond simple matching of CAD colors to pen colors. **AutoCAD** 2000, for example, provides a dizzying variety of options.

Revit, goes the opposite route, since it has preassigned colors to objects. A new Revit drawing preassigns named (not numbered) colors; the user can add additional named colors.

AutoCAD 2000 is an example of the fine control (some would say "to excess" and to the point of confusion) that today's CAD systems can have over the plotter. Provided the plotter (or printer) supports these features, you can specify, as shown by this illustration:

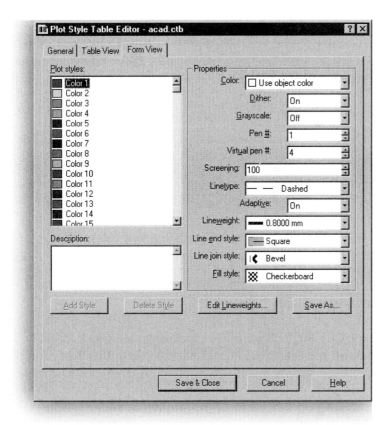

AutoCAD's Plot Style Table Editor

Dithering: Allows the plotter to approximate colors with dot patterns. Useful when the plotter supports fewer colors than does the CAD system. The drawback to dithering is that it can create the appearance of dotted linetypes on thin lines, and makes light colors less visible.

Grayscale: Converts colors to gray equivalents. This option is not usually necessary since most monochrome printers, such as laser printers, convert colors to gray on their own.

MICROSTATION PEN CONTROL

This discussion of color as it relates to plotting is not relevant to MicroStation, because lineweight, rather than color is used to control the appearance of a plot. If lineweight doesn't give you sufficient control, then you can write a pentables control plotter output based on any element characteristic you choose.

MicroStation controls the pen in a manner different from most other CAD packages. By default, the width of a plotted line is controlled by the line weight of the element. MicroStation plotting software can, however, alter the width of the plotted line based on many criteria, such as color, line style, or logical name of reference file.

Virtual Pen Number: Useful for specifying a virtual pen number between 1 and 255 for those nonpen plotters that simulate pen plotters. On the front panel of these nonpen plotters or via software, the pen numbers are mapped to pen width, fill pattern, end style, join style, and screening from the front panel on the plotter. This is similar in effect to AutoCAD's **Plot Style Table Editor.**

Screening: Determines the amount of ink the plotter places on the paper (0 = white; 100 = full color). This is useful for creating a screened plot.

Linetype: Overrides the object's linetype at plot time.

Adaptive Adjustment: Adjusts the scale of the linetype to complete the linetype pattern. While this creates a nicer looking linetype, it changes the linetype scale and should not be used if linetype scale is important.

Lineweight. Overrides the object's lineweight at plot time.

Line End Style. Specifies how the end of wide line should be drawn: Butt, Square, Round, and Diamond; overrides the object's line end style at plot time.

Line Join Style: Specifies how two lines are joined: Miter, Bevel, Round, and Diamond; overrides the object's line join style at plot time.

Fill Style: Specifies how objects are filled: Solid, Checkerboard, Crosshatch, Diamonds, Horizontal Bars, Slant Left, Slant Right, Square Dots, and Vertical Bar; overrides the object's fill style at plot time.

PLOTTERS FROM YESTERYEAR

In the first edition of this book, written in 1992, the electrostatic was the top-of-the-line plotter. The pen plotter was king; some firms were pushing thermal plotters as a cheaper, faster alternative; there were a few attempts to create large-format dot-matrix plotters; and the inkjet plotter didn't exist at all.

Today, almost everyone uses inkjet plotters; pens have gone extinct. No one uses thermal plotters. And electrostatic plotters used kerosene in the toner, which meant you had to get a toxic waste disposal company to remove the waste toner.

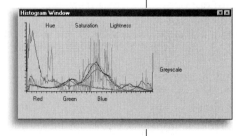

COLOR STANDARDS

Just as organizations have created standards for layers, they have also set standards for the use of colors in drawings.

AIA CAD Layer Guidelines

In its *CAD Layer Guidelines*, the AIA has specified a color for each layer. Examples of this are listed as follows. Notice the use of a color number, rather than a color name. The color numbers are based on Autodesk's ACI system:

Purpose	Layer Name	Color #	Color Name
Slab, new	S-SLAB-NEWW	11	Light Green
Slab, existing	S-SLAB-EXST	10	Light Yellow
Slab, to be demolished	S-SLAB-DEMO	7	White
Edge of slab, new	S-SLAB-EDGE-NEWW	3	Green
Edge of slab, existing	S-SLAB-EDGE-EXST	5	Blue
Edge of slab, to be demolished	S-SLAB-EDGE-DEMO	7	White
Slab reinforcing, new	S-SLAB-RBAR-NEWW	6	Magenta
Slab reinforcing, existing	S-SLAB-RBAR-EXST	5	Blue
Slab reinforcing, to be demolished	S-SLAB-RBAR-DEMO	2	Yellow

USCG Civil Engineering Technology Center

The United States Coast Guard Civil Engineering Technology Center provides another example of assigning object colors to plotter pens. They assigned the 255 combinations of pen widths and colors to MicroStation and AutoCAD drawings. The full list is in Appendix D; here are the first ten colors:

NCS Color #	Pen Plotter Pen width (mm)	Laser Plotter InkJet (in.)	Plot Color	MicroStation Color #	MicroStation Lineweight	AutoCAD Color #
1	0.18	0.007	Black	3	0	1
2	0.25	0.01	Black	4	1	2
3	0.35	0.014	Black	2	2	3
4	0.35	0.014	Black	7	2	4
5	0.5	0.02	Black	1	3	5
6	1	0.039	Black	5	7	6
7	1.4	0.055	Black	0	10	7
8	0.35	0.014	Halftone	9	2	8
9	2	0.079	Black	14	15	9
10	0.18	0.007	Black	10	0	10

RESOURCE

The United States Coast Guard Civil Engineering Technology Center at www.uscg.mil/mlclant/cetc/index.htm

Assigning Colors

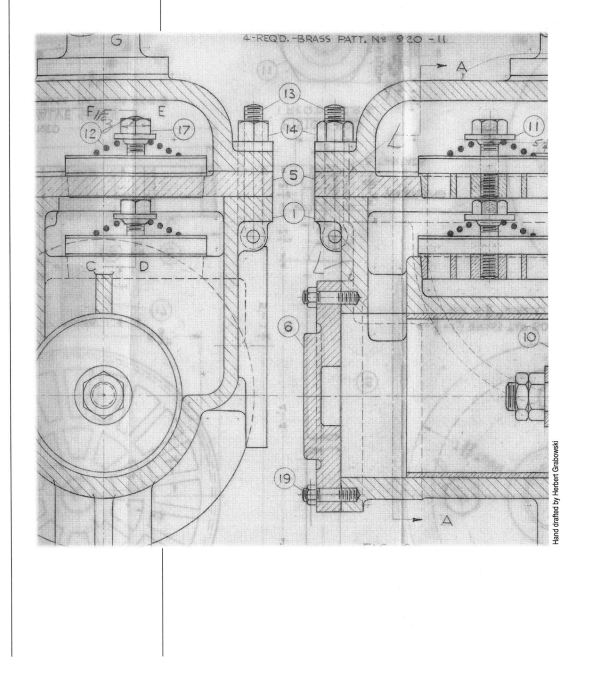

Fonts and Patterns, Linetypes and Widths   5

Traditionally, text in CAD drawings looked awful. In the early days, text was sometimes referred to as "stick fonts" due to their sparse look. The sparse look was a trade-off for display speed. An example is the Txt font, shown in the figure on the next page.

Faster computers meant that fonts could be smooth and filled. Examples are the commonly used Simplex font and the TrueType fonts (shown on the following page).

In addition to the fonts provided with the CAD software package, there are thousands of fonts available for CAD and most other software. During the 1990s, the price of fonts plummeted from several hundred dollars per font to practically free. For example, the CorelDraw package includes, at no extra cost, hundreds and hundreds fonts in PostScript and TrueType formats.

The drawback to rich font selection is that you might be tempted to use many of them, giving the "r*a*n*som* **n**o*te*" look to drawings. Several different fonts in a drawing looks unprofessional, and proves more difficult to read. Aim to work with two fonts in a couple of styles and sizes. One font should be a clear proportionally-spaced font used for notes and dimensions. The other font should be a monospace font, which is used to ensure that text in tables lines up.

CHAPTER SUMMARY
This chapter describes the standardized use of text fonts, linetypes, line widths, and hatch patterns in CAD drawings. The chapter includes some CAD standards.

TEXT STANDARDS

A text standard for your firm's CAD drawings need not be complex. Here is a suggested standard:

Font: Use the Simplex and Monotxt fonts.
Styles: Base the style names on size or other property.
Sizes: Work with these standard sizes:

Style	Height (in)	Height (mm)	Height (pts)
Small notes	3/32"	2.4 mm	7 pts *
Notes	1/8"	3.2 mm	9 pts
Subtitles	5/32"	4.0 mm	11 pts
Titles	7/32"	5.5 mm	16 pts

* 72.72 points = 1 inch.

Some common classes of CAD and TrueType font families.

CAD FONTS:

TXT font

Simplex (RomanS) font

Triplex (RomanT) font

Monospace font

Blueprint (handlettering font)

TRUETYPE (NON-CAD) FONTS:

Arial (sanserif font)

Times New Roman (serif font)

Sniff (distressed font)

Some firms use 3/32" as the standard text size for notes. But by going to the slightly larger size of 1/8", however, the text remains legible on drawings reduced to half-size.

> Small notes are 3/32" tall.
> Notes are 1/8" tall.
> Subtitles are 5/32" tall.
> Titles are 7/32" tall.

AIA and CSI Standards

Your clients may dictate other standards for fonts and styles. The following sections present CAD standards from several government organizations.

The AIA's *CAD Layer Guidelines* does not specify a standard for placing text in drawings, since its purpose is to describe layers. The layer structure does, however, specify the names of layers for placing text. For example, the A-ANNO-TEXT layer is meant for Architectural ANNOtation TEXT.

Layers that are used for text include:

AIA Layer Extension	Meaning
DIMS	Dimensions
IDEN	Identification tags
LABL	Text labels
NOTE	Notes, callouts, and key notes
NPLT	Nonplot information, construction lines
REV1	Drawing revision number 1
TAGG	Tags
TEXT	General notes and specifications
TTLB	Title block, sheet name or number

Neither the CSI's *MasterFormat* nor *Uniformat* specify layers for text.

> **TIP FONTS VERSUS STYLES**
> New AutoCAD users can confuse text *fonts* and *styles:*
> A **text font** defines the basic font. Text font names include *Txt*, *Simplex*, and *Times Roman*.
> A **text style** defines a variation on the font, specifying the height, oblique angle, and width. Text style names include *Cover*, *Title*, and *Notes*.
> To the Windows user, *12 pt bold* and *Arial underlined* are more familiar examples of style.

The Development of the CAD Font

It took two developments for CAD software to display better-looking fonts: (1) the hardware has to become fast enough; and (2) royalty-free TrueType fonts has to become a standard. Some CAD systems supported PostScript fonts until the royalty payments became an issue.

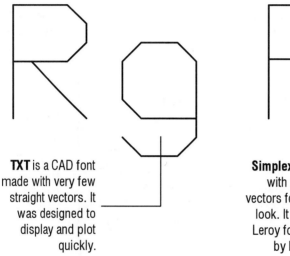

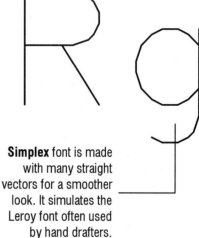

TXT is a CAD font made with very few straight vectors. It was designed to display and plot quickly.

Simplex font is made with many straight vectors for a smoother look. It simulates the Leroy font often used by hand drafters.

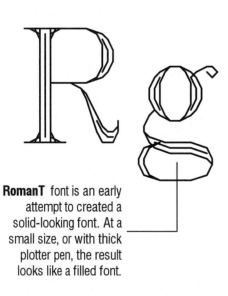

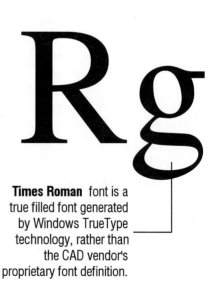

RomanT font is an early attempt to created a solid-looking font. At a small size, or with thick plotter pen, the result looks like a filled font.

Times Roman font is a true filled font generated by Windows TrueType technology, rather than the CAD vendor's proprietary font definition.

CalTrans Text Standard

CalTrans expects you to use one standard text font, Intergraph's #2 font, which is similar to the common Leroy font. For large titles, CalTrans suggests using Intergraph font #43, a filled sanserif block font.

CalTrans does not appear to specify a specific layer for text. The following fonts sizes are to be used in drawings:

Leroy Size	Imperial	Metric	1:50 Scale	Weight[1]	Font[2]	Description
120	0.12"	1.5mm	6.00'	0	2	Restrictive text areas
140	0.14"	1.8mm	7.00'	1	2	Notes
175	0.175"	2.2mm	8.75'	2	2	Subtitles
200	0.20"	3.5mm	10.00'	3	43	Titles
240	0.24"	3.0mm	12.00'	3	43	Sheet title
290	0.30"	3.7mm	14.50'	4	43	Project description

Notes:

[1] MicroStation font number. Note that CalTrans has its own font library based on the old IGDS (Intergraph) font library. CalTrans font 2 looks like MicroStation font 3; CalTrans font 43 is a homegrown file with a decent A-Z but incomplete set of lowercase characters. If a text standard for small to medium text uses MicroStation font 3, this is a good clue that it is using the standard MicroStation font library.

[2] MicroStation weight number.

TIP SPEEDING UP TEXT

For reasons of speed, CAD software provides numerous methods to reduce the redraw time of text. The methods vary, depending on the CAD package, but here are some common strategies:

Quick Text. Some CAD packages have an option that replaces text with a rectangular outline. This lets you see the location of the text but takes very little redraw time. The drawback is that the rectangular outline usually is a poor approximation of the line lengths.

Fill Off. CAD packages that can solid-fill text usually have an option to turn off the fill. Just before plotting, turn the fill back on.

Layer Off. Place text on their own layers, then turn off or freeze the text layers when not needed. This is the best option since it completely eliminates the text redraw time.

Fonts and Patterns, Linetypes and Widths

USCG Text Standard

The *Application and Standards Manual* written by the Civil Engineering Technology Center of the United States Coast Guard (USCG) specifies that only AutoCAD's RomanS font be used for all text on contract drawings; no other font is permitted. The USCG says that the RomanS font (a.k.a. Simplex) offers the following benefits:

- **Maximum Readability.** The clarity of this font provides the ability to plot readable text at a height of 0.1 inches.
- **Transportability.** Text entities are easily translated between CADD drawings.
- **Reduced-size Plotting.** The font and text sizes (listed as follows) provide clear full-size drawings and half-size reprints.

The USCG standard specifies that all text be placed on layer 20600.

The standard also specifies the default text height for each scale and for each discipline (architecture, metric, scientific, decimal, and engineering). These text heights are used for notes, dimensions, and other text in the drawing. The following table lists text heights for architectural and engineering units.

Architectural Scale	Text Height	Engineering Scale	Text Height
1/32"=1'	3' 2-3/8"	1"=5'	0' 6.0"
1/16"=1'	1' 7-3/16"	1"=10'	1' 0.0"
3/32"=1'	1' 0-13/16"	1"=20'	2' 0.0"
1/8"=1'	0' 9-5/8"	1"=30'	3' 0.0"
3/16"=1'	0' 6-3/8"	1"=40'	4' 0.0"
1/4"=1'	0' 4-13/16"	1"=50'	5' 0.0"
3/8"=1'	0' 3-3/16"	1"=60'	6' 0.0"
1/2"=1'	0' 2-3/8"	1"=80'	8' 0.0"
3/4"=1'	0' 1-5/8"	1"=100'	10' 0.0"
1"=1'	0' 1-3/16"	1"=200'	20' 0.0"
1-1/2"=1'	0' 0-13/16"	1"=250'	25' 0.0"
3"=1'	0' 0-3/8"	1"=300'	30' 0.0"
6"=1'	0' 0-3/16"	1"=400'	40' 0.0"
12"=1'	0' 0-1/8"	1"=500'	50' 0.0"

USACE Text Standard

The *CADD Details Library* published by the CADD/GIS Technology Center (hosted by the United States Army Corps of Engineers) recommends a specific font for two CAD programs:

CAD Package	Recommended Font
AutoCAD	RomanS (a.k.a. Simplex)
MicroStation	Font 1

The following layers are used for text:

Layer Name	Purpose
A-ANNO-DIMS	Witness and extension lines, dimension arrowheads, dots, slashes, and dimension text.
A-ANNO-NPLT	Construction lines, area calculations, and review comments.
A-ANNO-SYMB	Reference bubbles, matchlines and breaklines.
A-ANNO-TEXT	Detail title text, leaderlines, arrowheads and associated text, and notes.

The following text heights are to be used for drawings. The standard ensures that:

- **Text** is plotted 1/8 inch (3 mm) high.
- **Title text** is plotted 1/4 inch (6 mm) high.

IMPERIAL TEXT HEIGHTS

Detail Scale	Notes and Dimensions	Title Text
1/32 in. = 1 ft	4 ft	8 ft
1/16 in. = 1 ft	2 ft	4 ft
1/8 in. = 1 ft	1 ft	2 ft
1/4 in. = 1 ft	6 in.	1 ft
3/8 in. = 1 ft	4 in.	8 in.
1/2 in. = 1 ft	3 in.	6 in.
3/4 in. = 1 ft	2 in.	4 in.
1 in. = 1 ft	1-1/2 in.	3 in.
1-1/2 in. = 1 ft	1 in.	2 in.
3 in. = 1 ft	1/2 in.	1 in.
6 in. = 1 ft	1/4 in.	1/2 in.
Full Size (1=1)	1/8 in.	1/4 in.

METRIC TEXT HEIGHTS

Detail Scale	Notes and Dimensions	Title Text
1 : 200	600 mm	1200 mm
1 : 125	375 mm	750 mm
1 : 100	300 mm	600 mm
1 : 75	225 mm	450 mm
1 : 50	150 mm	300 mm
1 : 25	75 mm	150 mm
1 : 20	60 mm	120 mm
1 : 10	30 mm	60 mm
1 : 5	15 mm	30 mm
1 : 2.5	7.5 mm	15 mm
Full Size	3 mm	6 mm

TIP PLACE COMMENTS ON A NONPLOTTING LAYER

Drafters sometimes place notes they don't want plotted on a *nonplotting layer*, available in AutoCAD as of version 2000. The notes are visible during editing but do not plot. In versions of AutoCAD prior to 2000, use layer **DefPoints** for this purpose.

TSTC Text Standard

In contrast, the *A/E/C CAD Standard* published by the Tri-Service CADD/GIS Technology Center recommends the use of five fonts in drawings. They say that contrasting fonts delineates types of information, as follows:

- **Monotext** font for monospaced (vertically aligned) text, such as in schedules and some title blocks.
- **Proportional** font for general notes, labels, and most title blocks.
- **Slanted** font where text must distinguished from other text.
- **Filled** font for titles and cover sheets.
- **Outline** font is a substitute for filled fonts when plotting major titles to save plotting time.

Purpose	AutoCAD Font	MicroStation Font	TrueType Equivalent
Monotext	monotxt	font 3	...
Proportional	romans widthfactor = 0.8	font 1	...
Slanted	romans oblique angle 21.8d	font 23	...
Filled	swiss (truetype)	font 43	arialbd.tff
Outline	sasb	font 42	...

Text is placed on the following layers:

Layer Name	Purpose
**ANNO-DIMS	Witness/extension lines, dimension arrowheads/dots/slashes and dimension text.
**ANNO-KEYN	Keynotes with associated leader lines and arrowheads, ConDoc keynotes.
**ANNO-LEGN	Legends and schedules.
**ANNO-NOTE	General notes and remarks.
**ANNO-NPLT	Construction lines, reference targets, review comments, area calculations, and viewport windows.
**ANNO-REDL	Redlines, markups.
**ANNO-TEXT	Miscellaneous text and callouts with associated leader lines and arrowheads.

Note:
** represents the discipline code, such as A-, C-, and QY.

LINETYPES

Linetypes were used in hand drafting to differentiate views. For example, phantom and hidden lines show hidden parts. In some disciplines, linetypes denote data. For example, in a civil engineering drawing, a line broken by dots indicates a property border while a line broken by the letter T designates a telephone line: —T——T——T——T—.

Traditionally, CAD packages implemented rudimentary linetype capabilities, then let the feature languish. It took eight years for Autodesk to fix linetype spacing in polylines; MicroStation didn't allow the user to customize linetypes until Version 5.

Hardwired versus Customized Linetypes

CAD packages either supply *hardwired* linetypes or allow customized linetypes. Hardwire are linetypes that the user cannot change. Generic CADD and AutoSketch, for example, had hardwired linetypes. Up to Version 4, MicroStation had only hardwired linetypes. If your drawings needed a linetype different from the few supplied by these CAD packages, you had to be able to fake it or to do without.

Most CAD packages today allow you to create customized linetypes. They typically include a small collection of linetypes, and allow the user to add more linetypes.

One-Dimensional versus Two-Dimensional Linetypes

Most linetypes supplied by CAD packages are *one dimensional.* That means the linetype consists of lines, dots, and spaces. A few CAD packages support *two-dimensional* linetypes. This type includes text and shapes in the line.

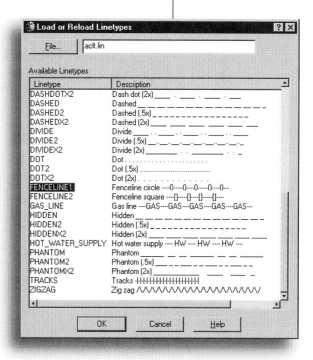

Some of the 1D and 2D linetypes provided with AutoCAD.

Software versus Hardware Linetypes

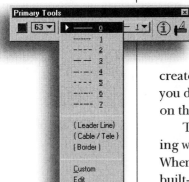

CAD software originally relied on the pen plotter to draw linetypes, since it was faster than the CAD package at generating linetypes. When the plotter draws the linetype, this is known as a *hardware* linetype; when the CAD package creates the linetype, this is known as a *software* linetype. Naturally, you don't use both the CAD software's and the plotter's linetypes on the same drawing.

To use hardware linetypes, you draw all lines in the CAD drawing with the continuous linetype. You differentiate lines by color. When it comes time to plot, you match CAD colors with the plotter's built-in linetypes.

Scaling Linetypes

When using linetypes, there are two scaling considerations: (1) the vector length, and (2) the plotted size.

When the CAD package draws the linetype, it begins at one end of the vector and draws the linetype pattern until it reaches the other end. Depending on the length of a vector (a line in the CAD drawing) and the scale of the linetype, the linetype might not show up or looks wrong (see following figure).

In MicroStation, the first eight linestyles (0 through 7, above) are hardwired. Additional linetypes are customizable in 1D and 2D formats, as shown by the dialog box below.

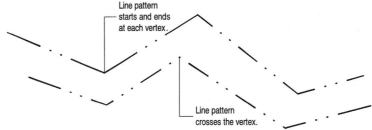

If the vector is too short and the linetype scaled too big, the linetype doesn't show up: the line appears solid. The reason is that the vector doesn't have enough length for the CAD software to draw the lines, gaps, and dots. The solution is to reduce the linetype scale or ignore the problem.

If the vector is long enough but the linetype scale is too small, the linetype also doesn't appear. In this case, however, the linetype makes its presence known: you see the line redrawn v-e-r-y slowly. This often happens when you are zoomed way out on a drawing.

Most CAD packages give you a choice of how to handle the linetype pattern at the two ends of the vector. The linetype starts at one end and stops in midpattern at the other end; the cleaner-looking alternative is to adjust the pattern at both ends to create a balanced linetype. The second choice is usually the default; in some drawings, you may want to turn off the feature.

Linetype Standards

The AIA's *CAD Layer Guidelines* specifies a linetype for each of its layers. It uses the AutoCAD names for linetypes. Most layers take the continuous linetype. The exceptions are:

- **Hidden**: Overhead items; overhang; access doors; equipment not in contract; foundation reinforcing; and slab reinforcing;
- **Center2**: Column grid; and centerlines.

Neither the CSI's *MasterFormat* nor *Uniformat* specify linetype standards.

USCG Linetype Standard

The *Application and Standards Manual* written by the Civil Engineering Technology Center of the United States Coast Guard specifies only that ISO linetypes should be used. The manual specifically mentions the ISO linetypes provided by AutoCAD. If necessary, linetypes can be placed on their own layer, as follows:

Layer Name	Purpose	Linetype Name
20100	Centerline	Center
20102	Centerline	Center3
20105	Centerline	Center2
20200	Phantom Line	Phantom
20202	Phantom Line	Phantom3
20205	Match Line	Phantom2
20300	Broken Line	Hidden
20302	Broken Line	Hidden3
20305	Broken Line	Dashed
20400	Solid Line	Continuous

TSTC Linetype Standard

The *A/E/C CAD Standard* published by the Tri-Service CADD/GIS Technology Center recommends the use of nine linetypes in drawings, as well as discipline-specific linetypes:

ID	Description	Example	MicroStation Linestyle	AutoCAD Linetype	Spacing [2]
0 [1]	Continuous	————	0	Continuous	
1	Dotted	. . . . . .	1	ISO07W100	0.5,-3
2 [1]	Dashed	– – – –	2	ISO02W100	12,-3
3	Dashed spaced	– – –	3	ISO03W100	12,-18
4	Dashed dotted	–.–.–.	4	ISO10W100	12,-3,.5,-3
6	Dashed double-dotted	–..–..–	6	ISO12W100	12,-3,.5,-3,0.5,-3
7 [1]	Chain	--- - --- -	7	ISO08W100	24,-3,6,-3
10	Dashed triple-dotted	–...–...	- [3]	ISO14W100	12,-3,.5,-3,.5,-3,.5,-3
11 [1]	Chain double-dashed	--- - - ---	- [3]	ISO09W100	24,-3,6,-3,6,-3

Notes:

[1] IDs 0, 2, 7, and 11 are included in International Standards Organization 128 (ISO 1982).
[2] Positive number signifies length of dash; negative number signifies length of space.
[3] This line style is not found in the default MicroStation line style resource file.

The TSTC apparently has linetype files for AutoCAD (tsaec.lin) and MicroStation (tsaec.rsc) at its Web site, but I have been unable to find them.

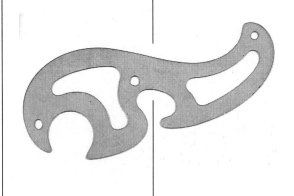

LINE WIDTHS

Line weights (line widths) help make a drawing easier to read by emphasizing important lines. While all CAD systems can specify line weight at plot time, the ability to show line widths on the screen varies according to the CAD software. MicroStation has been able to display line weights for most of its life, while AutoCAD only recently acquired the ability.

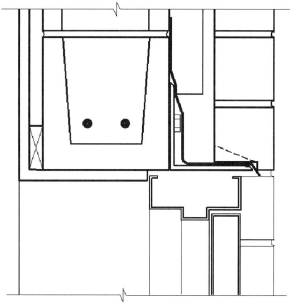

The effective use of line weights to improve the legibility of a detail drawing ("CADD Details Library: Report 1 - Architectural Details" by Tri-Service CADD/GIS Technology Center).

When line weights are specified at plot time, there is usually mapping involved. For example, you instruct the CAD software to plot all color #1 objects 0.1 mm wide. (Recall that CAD systems often refer to colors by number, not name.) For this reason, some standards specify a color number.

Linewidth Standards

The AIA's *CAD Layer Guidelines* does not specify line widths, nor does its layer structure specify them. Similarly, neither the CSI's *MasterFormat* nor *Uniformat* specify line width.

> **TIP MICROSTATION LINE WEIGHTS**
> The weight of MicroStation lines remains constant when plotted, no matter if the design is scaled up or down.

National CAD Standards Line Widths

In *Module 04*, the National CAD Standards organization recommends the following line widths in drawings:

Line Width	Metric (mm)	Name	Recommended Use
0.007"	0.18 mm	Fine	Material indications; surface marks; hatch lines.
0.10"	0.25 mm	Thin	1/8" (3 mm)-tall text; dimensioning and leaders; extension lines; break lines; hidden lines; dotted lines; dashed lines; setback line; centerline; grid line.
0.13".	0.35 mm	Med	5/32" (4 mm) to 3/8" (10 mm)-tall text; object lines; property lines; lettering; dimension tick marks.
0.020"	0.50 mm	Wide	7/32"(6 mm) to 3/8" (10 mm)-tall text; edges of interior and exterior elevations; profiling; cut lines; property lines; section cutting plane line.
0.028"	0.70 mm	X-Wide	1/2" (13 mm) to 1" (25 mm)-tall text; match lines; drawing border.

TSTC Line Width Standard

The *A/E/C CADD Standard* published by the Tri-Service CADD/GIS Technology Center recommends that just five line widths be used. The typical use for each line weight is:

- **Fine** (0.18 mm) for hatch lines; sparingly used since it does not reproduce well.
- **Thin** (0.25 mm) for dimension lines, leaders, note leader lines, line terminators (arrowheads, dots, slashes), phantom lines, hidden lines, centerlines, long break lines, schedule grid lines, and object lines seen at a distance.
- **Medium** (0.35 mm) for minor object lines, dimension text, text for notes/callouts, and schedule text.
- **Wide** (0.50 mm) for major object lines, cut lines, section cutting plane lines, and titles.
- **Extra wide** (0.70 mm) for minor title underlining, lines requiring special emphasis. For very large scale details drawn at 3 in. = 1 ft-0 in. or larger, the extra wide width should be used for the object lines. Extra wide widths are also appropriate for use as an

---

> **TIP LINE WEIGHTS AND COLORS**
>
> In MicroStation, it is redundant to define the width of plotted lines by both color and line weight. MicroStation line weights are used for screen display only; pen tables use color to control the width of a plotted line.
>
> Using both color and line weight makes sense, however, when files must be translatable between MicroStation and AutoCAD formats. Versions of AutoCAD R14 and earlier cannot display nor plot based on line widths, so color is used instead.

elevation grade line, building footprint, or top of grade lines on section/foundation details.
- **Option 1** (1.00 mm) for major title underlining and separating portions of drawings.
- **Option 2** (1.40 mm) for border sheet outlines and cover sheet line work, and as an option for the designer as required.
- **Option 3** (2.00 mm) border sheet outlines and cover sheet line work and as an option for the designer as required.

Line Thickness	Pen Designation	mm	in.	MicroStation Line Weight
Fine	0000	0.18	0.007	wt = 0
Thin	000	0.25	0.010	wt = 1
Medium	0	0.35	0.014	wt = 2
Wide	1	0.50	0.020	wt = 3
Extra Wide	2.5	0.70	0.028	wt = 5
Option 1	3.5	1.00	0.040	wt = 7
Option 2	n/a	1.40	0.055	wt = 10
Option 3	n/a	2.00	0.079	wt = 15

USACE Line Weight Standard

The *CADD Details Library* published by the CADD/GIS Technology Center (hosted by the United States Army Corps of Engineers) recommends the following line widths for drawings. The USACE comments that "the line weights shown in the table beside each color have to be associated with that color in order to ensure that details are plotted at the correct visual weight."

Color	AutoCAD Color #	MicroStation Color #	Line/Pen Width	MicroStation Weight
Blue	5	1	0.007 in. (0.18 mm)	LW = 0
Grey	8	9	0.007 in. (0.18 mm)	LW = 0
Red	1	3	0.010 in. (0.25 mm)	LW = 1
Green	3	2	0.010 in. (0.25 mm)	LW = 1
Yellow	2	4	0.014 in. (0.35 mm)	LW = 2
Magenta	6	5	0.014 in. (0.35 mm)	LW = 2
Cyan	4	7	0.020 in. (0.50 mm)	LW = 3
White	7	0	0.028 in. (0.70 mm)	LW = 5

The nib of a 0000-wide technical pen. These pens were the pride of hand drafters in the 1980s, but required much care and cleaning.

HATCH PATTERNS

Hatch patterns are much like linetypes. A collection of patterns are supplied with every CAD package; hatch patterns are never hard-coded into the software, as linetypes sometimes are. The software documentation usually describes how to create your own custom hatch patterns, or you can purchase commercial collections at a typical price of about $1 per pattern — much cheaper than the time it would take to write your own.

Hatch patterns are like codes that visually describe the properties of the objects in the drawing. Every discipline has its own set of hatch patterns with unique meaning. Thus, there tends not to be an office standard for hatch patterns; rather, each CAD operator refers to the patterning prescribed by the discipline, whether architecture, mechanical design, or mapping.

Hatch Pattern Standards

There seem to be no published standards for hatch patterns in the literature I researched. This could be because disciplines, such as

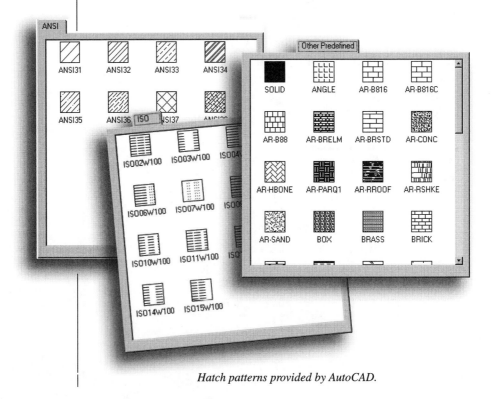

Hatch patterns provided by AutoCAD.

mechanical engineers and mapmakers, have their own standards for hatch patterns.

The AIA's *CAD Layer Guidelines* does not specify hatching, since its purpose is layers. The layer structure does, however, specify that hatching be placed on layers ending with:

AIA Layer Suffix	Purpose
-PATT	Hatch pattern
-SHAD	Solid shading
-HTCH	Hatching

For example, layer L-DETL-PATT-DEMO is for textures and hatch patterns that show landscaping details to be removed.

Neither the CSI's *MasterFormat* nor *Uniformat* specify hatching, nor layers for hatches.

The United States Coast Guard's adaptation of *MasterFormat* specifies that crosshatching be placed on layer 20900.

RESOURCE

Sixty-four free hatch patterns for AutoCAD can be downloaded from the Web site at www.hatchpatterns.com

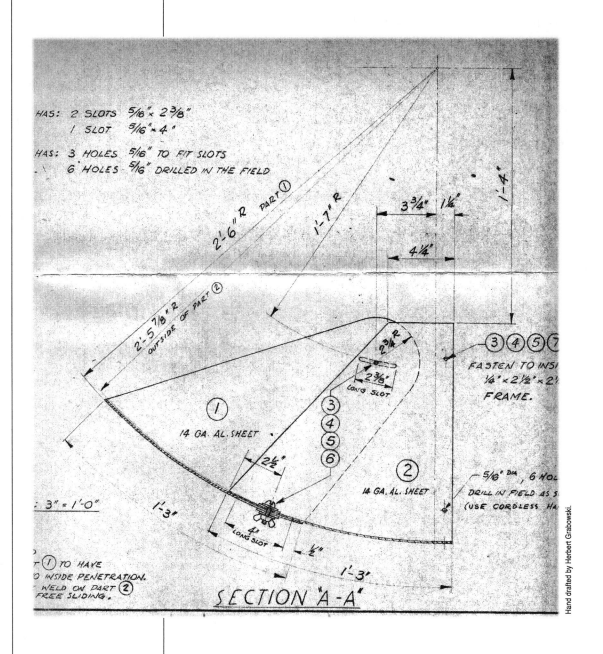

Scale Factors and Dimensions

6

A difficulty with CAD is that you usually work with two different scale factors: (1) most of the drawing is drawn full size at 1:1 scale; and (2) parts of the drawing are drawn at a second scale factor that reflects the final plot size.

Elements such as text, linetypes, dimensions, and hatch patterns cannot be drawn full size. Instead, they must be drawn at a size that makes them legible when plotted. For example, text on a blueprint should be about $1/8$" high, but you cannot draw text at $1/8$" tall on a 1:1000 drawing because it would plot as a series of tiny, illegible dots. There are two solutions:

Solution 1: Determine the scale factor before the first text, linetype, and hatch pattern is drawn.
Solution 2: Draw these scale-dependent objects in paper space (a.k.a. layout mode).

CHAPTER SUMMARY
In this chapter, you learn about scaling drawing elements, such as text and hatch patterns, to be legible once plotted.

This chapter also describes dimensioning standards for CAD drawings, possibly the most complex aspect of CAD. Dimensions involve seemingly endless options, particularly since most CAD packages include multiple, predefined dimension styles. This makes it possible for you to use standards, such as ANSI and ISO, which is particularly important if your firm does international work.

SCALE FACTORS

In many drawings, you use both solutions. Some text, linetypes, and hatch patterns are typically drawn in model space. These must be scaled up (usually) to make them larger. When plotted, they appear the correct size.

Notes, title blocks, and dimensions are typically drawn in paper space. In paper space, the model is scaled down to plot size. Scale-dependent objects such as text, therefore, can be drawn full-size.

In summary:

Objects	**Model Space**	**Paper Space**
Scale-independent		
Lines, circles, surfaces, solids, etc.	Full size	Scaled smaller *
Scale-dependent		
Text, patterns, linetypes, dimensions	Scaled larger	Full size

Note:
* Most, but not all, drawings are scaled to make them small enough to fit the plot media.

Determining the Scale Factor

Calculating the scale factor is not difficult; setting it in the CAD drawing is perhaps a bit more difficult due to the confusing method by which CAD software implements scale factors. For example, AutoCAD does not have one global scale factor; instead, you set scale factors independently for text, dimensions, linetypes, and hatch patterns.

The scale factor is usually determined by making normal text appear $\frac{1}{8}$" tall. The scale factor is based on the drawing's plot scale.

The scale factor can be set automatically via a setup routine (such as AutoCAD's **MvSetup** command) or manually. When performed manually, you can use a table such as the one that follows, or calculate the scale factor using a formula, shown later.

Plot Scale	Scale Factor	1/8" Text Height
1/8" = 1'-0"	1 : 96	12"
3/16" = 1'-0"	1 : 64	8"
1/4" = 1'-0"	1 : 48	6"
1/2" = 1'-0"	1 : 24	3"
1" = 1'-0"	1 : 12	1.5"
1 1/2" = 1'-0"	1 : 8	1"
1' = 1000'	1 : 1000	10' 5"
1' = 500'	1 : 500	5' 2.5"
1' = 250'	1 : 250	2' 7.25"
1' = 100'	1 : 100	1' 0.5"
1' = 50'	1 : 50	6.25"
1' = 10'	1 : 10	1.25"
1' = 2'	1 : 2	0.25"
1' = 1'	1 : 1	0.125"
2' = 1'	1 : 0.5	0.0625"

Text height (in inches) can be calculated using the following formula:

$$\text{Text Height} = \frac{1/8" \text{ Size}}{\text{Scale Factor}}$$

For example, the text height (in inches) for a drawing plotted at 1:50 scale is calculated, as follows:

$$\text{Text Height} = \frac{1/8"}{1:50}$$

which equals

$$6.25" = \frac{0.125"}{0.020}$$

With the text scale determined for the drawing, the same factor is applied to dimension text, leader text, hatch patterns, and linetypes.

DIMENSIONING WITH CAD

Dimensions are probably the most complex aspect of CAD. Although the typical dimension has only four components (extension line, dimension line, arrowhead, and dimension value), each can be expressed in a wide variety of styles. Disciplines and regions each have their own way of drawing dimensions.

For this reason, almost all CAD packages provide a way to customize the look of dimensions, called the *dimension style*.

CAD packages offer different kinds of dimensioning and, naturally enough, give them different names. Here is a list of the ways that a CAD system can dimension.

- **Manual dimensioning**: you draw all parts of the dimension yourself.
- **Semiautomatic dimensioning**: you tell the CAD package where the two extension lines go and the location of the dimension line; the CAD software draws the dimension.
- **Automatic**: you select an object and the CAD package dimensions it.
- **Nonassociative dimensioning:** after the dimension is drawn, it is independent of the object it measures.
- **Semiassociative dimensioning:** when the *dimension* is stretched or condensed, it automatically updates its value.
- **True associative dimensioning:** when the object is stretched or condensed, the dimension is automatically updated.

When given the choice in a CAD package, select the automatic, true associative dimensions.

Some CAD packages take automatic dimensioning one step further. A CAD package once sent to me from Moscow was able to dimension every feature in the drawing with a single command! This is an interesting approach, since it easier to erase a dimension than it is to place it.

> **DIMENSIONS IN 3D**
> Use caution when dimensioning 3D objects. Is the distance between two points measured parallel to the screen, or is it the shortest distance?

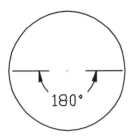
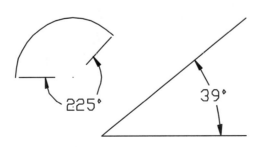

A variety of radial dimensions.

The Anatomy of a Dimension

To better understand why dimension styles are needed, let's look at their complexity. Each type of dimension typically consists of four elements:

· Dimension line
· Extension line
· Arrowhead
· Dimension text

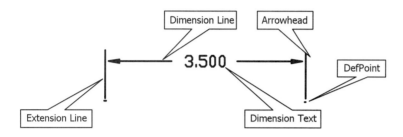

But each part has many variables. For example, the arrowhead might be an arrow, a dot, a slash, or nothing at all. The arrowhead's size, color, and layer can differ from the other dimension elements.

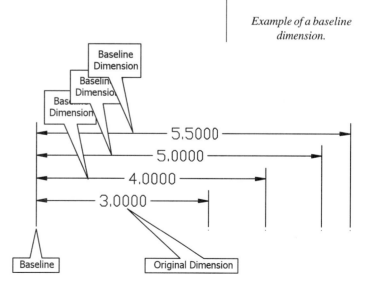

Example of a baseline dimension.

There are these basic types of dimension:

· Linear dimension
· Aligned dimension
· Radial dimension
· Leader

Other types of dimensions include baseline, continued, data tables, match lines, call-outs, break lines, labeling coordinates, tolerances, coordinates systems, and clearances.

Dimensioning Standards

Many CAD packages include presets for several international dimensioning standards. The most common are:

- **ANSI** (American National Standards Institute) is often the default for CAD drawings that use imperial units.
- **DIN** (Deutsche Industrie Norm) defines the standards for dimensioning German drawings.
- **JIS** (Japanese Industrial Standard) defines dimension standards for Japanese drawings.
- **ISO** (International Organization of Standards) defines the dimension standard for metric drawings.
- **Geometric tolerancing** is a type of dimensioning that defines how surfaces should be machined. There are two standards: the imperial ASME Y14.5 and the metric ISO 1101.

A CAD package implements a particular dimension standard via *system variables*. AutoCAD, for example, has 67 dimension variables that affect the look of a dimension. By setting the variables to appropriate values, the CAD package mimics the dimension standard, whether ANSI, ISO, DIN, JIS, or a custom standard.

The following table lists the AutoCAD system variables that differ, depending on the selected dimension standard.

MicroStation controls dimension style through the **Dimensions Settings** dialog box. Predefined dimension styles are accessed by selecting **Settings | Manage** menu, which brings up the **Select Settings** dialog box. The available dimensioning styles are:

ANSI ANSI Y14.5 Mechanical
AS1100 Australian Mechanical
DIN Mechanical
ISO ISO Mechanical
JIS JIS Mechanical
GraphStds Architectural Graphics Standards
PArch Intergraph Project Architect

Dimension Variable	Name	ANSI	DIN	JIS	ISO
Arrow size	DimASz	0.1800	2.5000	2.5000	2.5000
Center mark size	DimCen	0.0900	2.500	0.0000	2.500
Dimension line spacing	DimDLi	0.3800	3.7500	7.0000	3.7500
Extension above dimension line	DimExe	0.1800	1.2500	1.0000	1.2500
Extension line origin offset	DimExO	0.0625	0.0625	1.0000	0.0625
Gap from dimension line to text	DimGap	0.0900	0.6250	0.0000	0.6250
Suppress outside dimension lines	DimSOXD	Off	Off	On	Off
Force line inside extension lines	DimTOFL	Off	On	On	On
Linear unit format	DimLUnit	Decimal	Windows System	Decimal	Windows System
Alternate units	DimAltU	Decimal	Windows System	Decimal	Windows System
Scale factor, alternate units	DimAltF	25.4000	0.0394	0.0394	0.0394
Decimal separator	DimDSep	.	,	.	,
Decimal places, alternate units	DimAltD	2	2	2	4
Decimal places, alternate tolerance	DimAltTD	2	2	2	4
Place text above the dimension line	DimTAD	Centered	Above	Above	Above
Text inside extensions is horizontal	DimTIH	On	Off	Off	Off
Text outside horizontal	DimTOH	On	Off	Off	Off
Text height	DimTxt	0.1800	2.5000	2.5000	2.5000
Zero suppression	DimZIN	All zeros	Trailing zeros	Trailing zeros	Trailing zeros

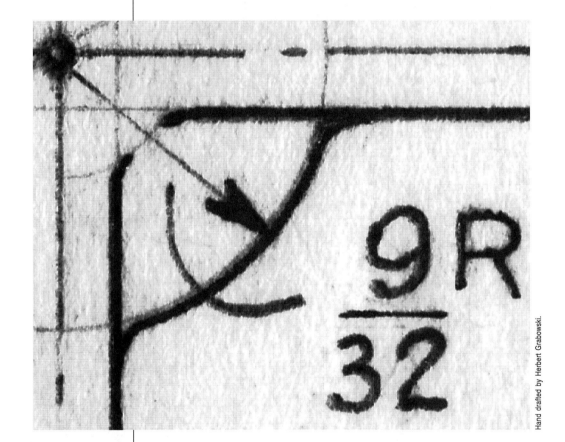
Hand drafted by Herbert Grabowski.

Standard Drawings and Templates

7

Just as there are international standards for dimensions, there are standards for drawing size, borders, and title blocks. In addition, there is a standard for placing multiple details on a single drawing. The international standards are known as:

- **ANSI** American (imperial).
- **DIN** German (metric).
- **ISO** International (metric).
- **JIS** Japanese (metric).

Some CAD packages include drawings (called *templates*) based on these standards.

CHAPTER SUMMARY

In this chapter, you learn how to incorporate standards in a template drawing. You also learn about standard title blocks and drawing borders, as well as in which order to place drawing sheets.

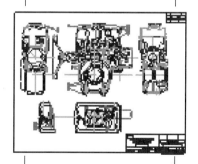

DRAWING SIZES

When you create a drawing, you draw it full size. Being digital, the size of the drawing doesn't matter. (There is one exception: some CAD packages constrain the maximum drawing size, such as MicroStation up through version J.)

When it comes time to plot the drawing on paper, you are usually constrained to a half-dozen sizes of media, ranging in size from under a foot to about 4 feet wide.

In imperial measurements, the sizes are called **A** (the smallest) through **E** (the largest). There are additional special purpose sizes, such as the extra-long **J**-size used by the aircraft industry.

In the metric world, the sizes are called **A0** (the largest) through **A4** (the smallest).

While there are variations of each size (engineering, architectural, and metric), they all have one thing in common: each larger size doubles the paper area. For example, C-size (17" x 22") is twice the area of B-size (11" x 17").

<u>ANSI</u>	<u>Engineering</u>	<u>Architectural</u>	<u>Nearest Metric Equivalent</u>	
Standard sizes				
A (Letter)	8.5 x 11"	9 x 12"	A4	210 x 297 mm
B	11 x 17"	12 x 18"	A3	297 x 420 mm
C	17 x 22"	18 x 24"	A2	420 x 594 mm
D	22 x 34"	24 x 36"	A1	594 x 841 mm
E	34 x 44"	36 x 48"	A0	841 x 1189 mm
Other sizes				
F	28 x 40"	30 x 42"	...	
Legal	8.5 x 14"	...	A5	148 x 210 mm
Executive	7.24 x 10.5"	...	B5	182 x 257 mm

ISO (Metric) Sheet Sizes: A0 through A4

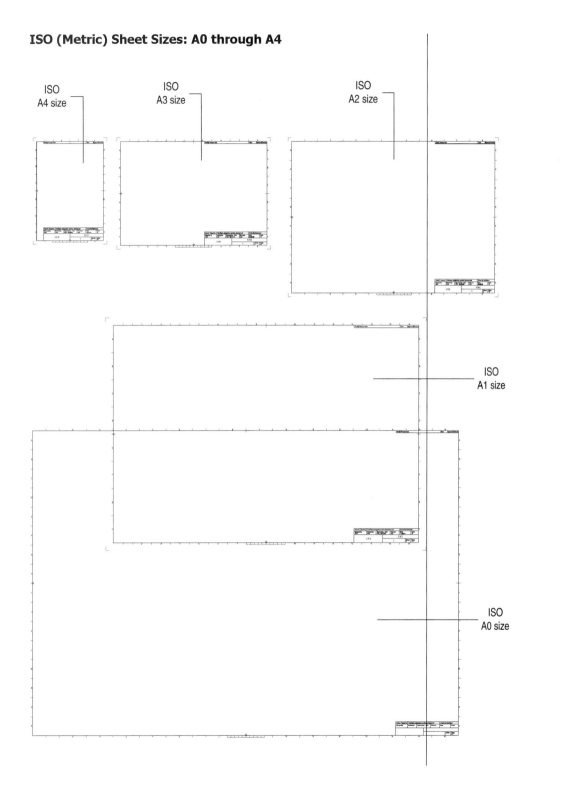

Standard Drawings and Templates

ANSI Sheet Sizes: A through E

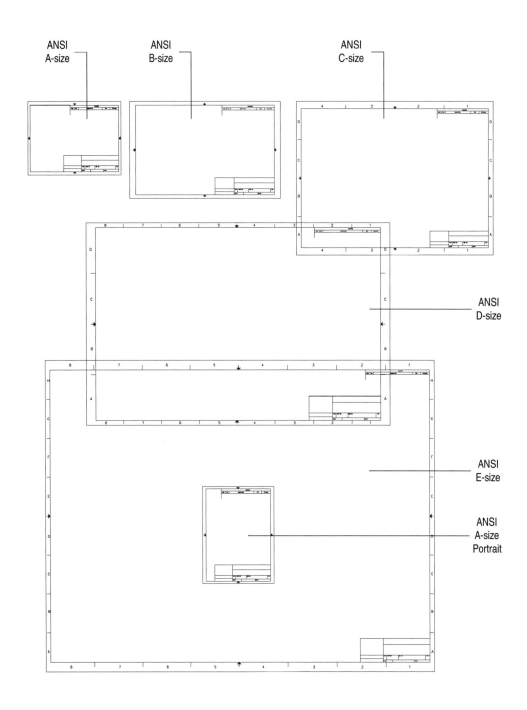

JIS (Japanese) Sheet Sizes: A0 through A4

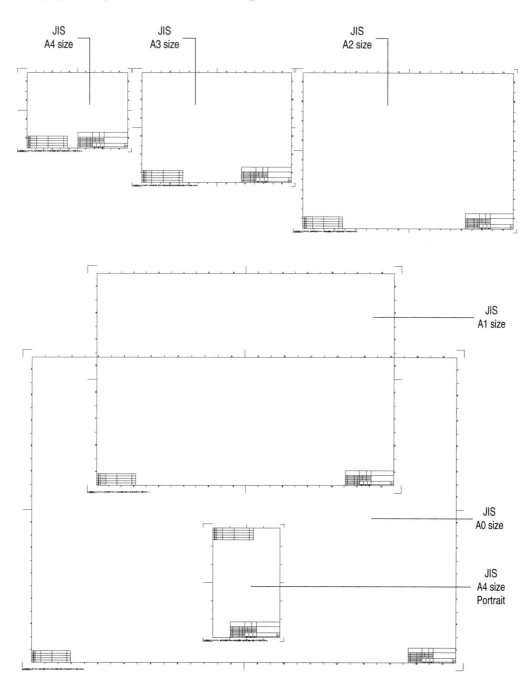

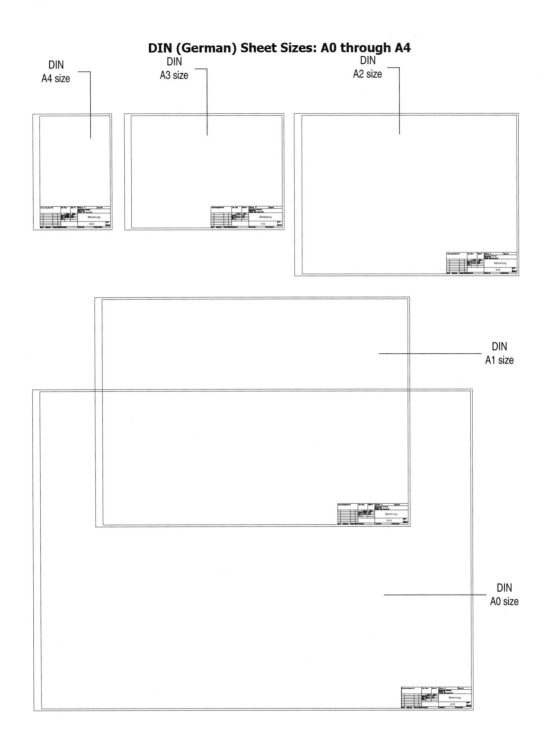

BORDERS AND TITLE BLOCKS

Drafting uses a border around drawings to help organize them. The border consists of these parts:

Title Block. Describes the name/number of the drawing and everyone involved, including the drafter, the checkers, owner, and client. Usually located in the lower-right corner, or along the right edge so that it can be read without needing to unfold or unroll the drawing.

Drawing Area. Contains one or more drawings, key plans, and schedules.

Fold Indicators. Triangles or arrows that indicate where the drawing is to be folded.

Revision List. Lists the changes made to the drawing, along with who and when. Sometimes the parts list appears here.

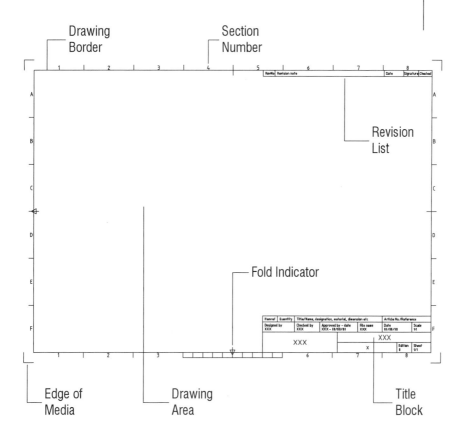

Standard Drawings and Templates

Edge of Media: Printers and plotters cannot print to four edges of the media. They require a small strip along the edges to grip the media, called the *margin.* Many plotters include an option called "expanded plot size" or "maximum print area," which shrinks the width of the margin at the risk of smudging one or more edges of the drawing.

Drawing Modules: Refers to a detail portion of the drawing, using a row and column description of letters and numbers, such as A1 in the upper right corner of metric drawings.

CSI Drawing Sheet Specification

The Construction Specifications Institute specifies that the drawing sheet be organized a bit differently:

Title Block. Along the right edge.

Drawing Modules. A1 is at the lower-right corner.

Notes Block. If notes are required, they are placed in a block in column 5. From top to bottom, the notes block contains: General Sheet Notes, Reference Keynotes, and Sheet Keynotes.

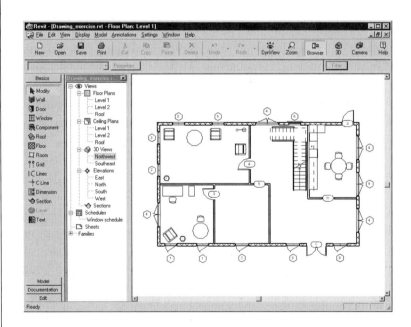

Model view displayed by Revit. Notice that only the model is displayed.

PLACING SHEETS IN ORDER

In chapter 2, "Naming Drawings and Creating Symbols," you learned about standards for giving file names to CAD drawings. The final output from drafting, however, is not the CAD drawing but the *plotted sheet.*

In many cases, the sheet consists of a drawing surrounded by the border and title block. In some cases, there may be several drawings on a sheet.

CAD software makes a distinction between model mode and sheet mode. *Model mode* is where you create the drawing full size; *sheet mode* is where the drawing is scaled to fit the output media. MicroStation use the term "sheet," while AutoCAD calls these two modes "model" and "layout" (prior to AutoCAD 2000, layout mode was called "paper space").

There are, hence, two naming systems: one for drawing file names (chapter 2), and another for sheet files (in this chapter).

The Single Design Model

Sheet mode becomes more important with CAD programs that work with a single model, such as most mechanical and architectural design software. Examples include SolidWorks, Alventive

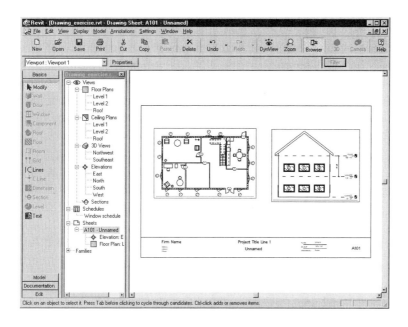

Sheet view displayed by Revit. Notice that this sheet view includes two model views and the drawing border.

IronCAD, Revit, Bentley Architectural TriForma, Bricsnet Architectural, and AutoCAD Architectural Desktop.

In all these programs, you design the entire model (whether a mechanical part or an entire building). Then, 2D plans are taken of the model, using the standard views — front, top, side — as well as isometric and slices through the model.

Standards for Sheet Identification

The Construction Specification Institute's *Uniform Drawing System* specifies the order in which construction drawings appear in a set. The UDS uses five characters — AANNN — two letters and three digits, such as CS101 for the General Site Plan in the Civil-Survey section.

A	Discipline designator
A**A**	Modifier
AA**N**	Sheet type designator; ranges from 0 to 9
AA**NN**	Sequence numbers

The sheets are not bundled together in alphabetical order, but in the order listed in the following table. The **G** section (general) comes before **H** (hazardous materials), while the last section is **R**, resources.

This nonalphabetical order is a problem for computers, since they sort file names alphabetically, rather than logically.

Each engineering discipline is likely to have drawings stored in their own separate folder (or directory) with read/write permission for their discipline, and read-only for other disciplines. On a large project, each discipline might be in a different geographic location, or even at a different company.

Each discipline is usually responsible for their own plotting and collating. The real problem occurs for the person coordinating the disciplines; this coordinator must gather the drawings from all disciplines and assemble them into the sets issued to contractors.

In addition to the Uniform Drawing System identification, drawings are given a page number, such as 001, showing their location in the drawing set. Assigning page numbers and creating the index sheet (sometimes with last-minute revisions) can be a logistical problem.

Designator and Description

G   General
- GI   Informational
- GC   Contractual
- GR   Resource

H   Hazardous materials
- HA   Asbestos
- HC   Chemicals
- HL   Lead
- HP   PCB
- HR   Refrigerants

C   Civil
- CD   Demolition
- CS   Survey
- CG   Grading
- CP   Paving
- CI   Improvements
- CT   Transportation
- CU   Utilities

L   Landscape
- LD   Demolition
- LI   Irrigation
- LP   Planting

S   Structural
- SD   Demolition
- SS   Site
- SB   Substructure
- SF   Framing

Designator and Description

A   Architectural
- AS   Site
- AD   Demolition
- AE   Elements
- AI   Interiors
- AF   Finishes
- AG   Graphics

I   Interiors
- ID   Demolition
- IN   Design
- IF   Furnishings
- IG   Graphics

Q   eQuipment
- QA   Athletic
- QB   Bank
- QC   Dry Cleaning
- QD   Detention
- QE   Educational
- QF   Food Service
- QH   Hospital
- QL   Laboratory
- QM   Maintenance
- QP   Parking Lot
- QR   Retail
- QS   Site
- QT   Theatrical
- QV   Video/Photographic
- QY   Security

F   Fire Protection
- FA   Fire Detection and Alarm
- FX   Fire Suppression

Designator and Description

P   Plumbing
- PS   Plumbing Site
- PD   Process/Plumbing Demolition
- PP   Process Piping
- PQ   Process Systems
- PE   Process Electrical
- PI   Process Instrumentation
- PL   Plumbing

M   Mechanical
- MS   Site
- MD   Demolition
- MH   HVAC
- MP   Piping
- MI   Instrumentation

E   Electrical
- ES   Site
- ED   Demolition
- EP   Power
- EL   Lighting
- EI   Instrumentation
- ET   Telecommunications
- EY   Auxiliary Systems

T   Telecommunications
- TA   Audio Visual
- TC   Clock and Program
- TI   Intercom
- TM   Monitoring
- TN   Data Networks
- TT   Telephone
- TY   Security

R   Resource
- RC   Civil
- RS   Structural
- RA   Architectural
- RM   Mechanical
- RE   Electrical

NOMENCLATURE

Template is the word used commonly in the Windows world, and has been adopted by some CAD packages. AutoCAD used to call it *prototype*, while MicroStation calls it a *seed* drawing.

CREATING A TEMPLATE DRAWING

In the previous chapters, you have been reading about how standards apply to different parts of the CAD drawing — layers, colors, symbols, file names, line patterns and widths, text, dimensions, and scaling. There are two ways to bring all this information together into a single reference:

1. Embed the standards in a template drawing.
2. Write up the standards in a reference book.

In the remainder of this chapter, you learn about creating template drawings with a number of CAD packages. In the chapter following, you create the CAD standard reference book.

What is a Template Drawing?

A *template drawing* eliminates the tedium the CAD operator faces when setting up each new drawing to your firm's (or your client's) CAD standards. The template drawing is very similar to a Word template file: both predefine styles, paper sizes, and even elements common to every drawing, such as the title block and border. A template drawing is like a catalyst: it helps create new drawings, but is not consumed in the process.

When your firm has just one drawing type, say all your drawings are vertical A-size with no border (such as for technical illustrations), then you need to create just one template. When your firm is multidisciplinary and/or work with many different clients, then you'll need to create many template files — one for each kind of drawing.

Here are examples of some template drawings you might need to create:

· Imperial and metric dimensions.
· Sheet size, such as A, A vertical, B, C, D, E, and custom sizes.
· Two- and three-dimensional drawings.
· Architectural, structural, mechanical, landscape, HVAC, P&ID, path, and electrical schematics.
· Your firm's standards, and your client's standards.

Sometimes the number of different template drawings depends on the CAD system. For example, MicroStation requires that you create separate seed files for 2D-only and 3D drawings. AutoCAD

LT has poor support for 3D and discipline-specific add-ons, so there's little point in creating 3D and vertical templates.

Preparing the Template Drawing

There are a large number of steps to preparing a template drawing. Once you complete these steps for one template drawing, you don't ever need to go through them again — at least for new drawings based on this particular template.

To reduce the amount of effort in creating additional template drawings, you can base them on the first template drawing. For example, all that changes on additional template drawings may be the drawing border and scale factors.

The CAD program might be able to assist you in creating the template drawing. Some include template drawings that you can use directly, or modify to your needs. Some also include a command or dialog box that helps you set many of the options for the template drawing.

AutoCAD

TurboCAD

Revit

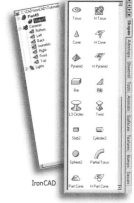

IronCAD

TIP OVERVIEW OF DRAWING SETTINGS

A number of CAD packages come with an overview window that shows the current setting of many options. Examples are illustrated here. The name given to this window, however, varies with the software package, and sometimes you need to open more than one window:

AutoCAD: The DesignCenter and the Properties windows.
Revit: The Project browser
TurboCAD: The Properties Palette.
IronCAD: The Catalog Browser and the Scene Browser.

Finally, check with your industry association or local CAD user group. They may already have created template drawings that you can use or modify. For example, the CADD/GIS Technology Center has predefined workspaces for MicroStation users at tsc.wes.army.mil/Products/standards/workspace/default.asp.

Template Creation Steps

Here are the steps you should go through to set up a template drawing:

Step 1: Open a new drawing file. This forms the basis of the template drawing. If you have the option, you may find it easier to start with an existing template drawing. AutoCAD, for example, includes dozens of template drawings in a variety of sizes and with international standards.

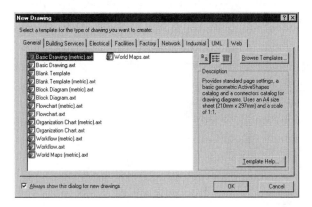

Actrix lets you start with one of many predefined template drawings included with the package.

Step 2: Set the unit of measurement. The options depend on the capabilities of the CAD package. Most provide architectural units (feet, inches, and fractional inches) and metric (also known as decimal units.) Common units include:

- Feet and fractional inches: 12'-3/4"
- Feet and decimal inches: 12.75'
- Fractional or decimal inches without feet: 144-3/4" or 144.75"
- Metric or decimal units: 12.75
- Scientific units: $1.275E+01$

> **TIP AZIMUTH**
> *Azimuth* refers to the angle measured clockwise from North.

Note that you will be doing dimensioning in layout mode (on the sheet file).

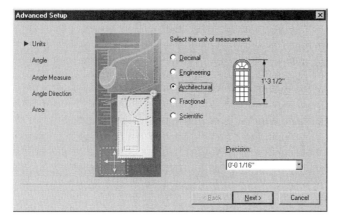

AutoCAD includes a "wizard" that steps you through some, but not all, the steps for setting up the parameters of a new drawing. Here the wizard asks for the units of measurement.

Step 3: Specify the resolution of measurement. Most CAD systems have accuracy levels that are extreme and unusable in the real world; nobody makes a measurement to 14 decimal places. Instead, you tend to measure to the nearest $\frac{1}{4}$" (in architecture and construction) or to the nearest hundredths of a millimeter (in machine tooling) or the nearest light-year (in Star Trek navigation maps).

- For fractional units, set the resolution to the nearest $\frac{1}{4}$" or whatever resolution your discipline typically employs.
- For metric, decimal inches, and scientific notation, set a limit to the resolution in decimal places, such as two place s (i.e. 0.01).

Step 4: Select the style of angular measurement. Angles are measured in a variety of ways. Depending on your discipline, and on the abilities of the CAD program, you may need to select:

- **System of measurement:** Decimal degrees (most common; 12.75°), degree-minutes-seconds (12° 45'00"), grad (12 g), radian (1.57 r), and surveyor's units, also called "bearings" (N 12° 45'00" E or S 12° 45'00" W).
- **Direction of 0 degrees:** North (most common for civil engineering and surveying), South, East (most common for other disci-

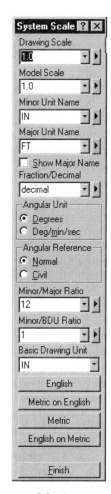

Selecting measurement styles in CADMAX.

Standard Drawings and Templates 119

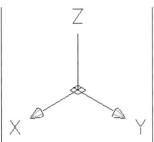

AutoCAD's xyz-icon (called the "UCS icon") indicates the origin of the drawing, as well as the positive direction of the three axes.

plines), West, or some arbitrary angle between 0 and 360 degrees.

- **Direction of angle measurement:** Clockwise or counterclockwise (most common).

Step 5: Set the drawing origin and extents. The *origin* is at 0,0,0 where the x, y, and z axes meet. Sometimes the origin needs to be relocated, such as for maps.

The *limits* define the extents of the drawing; some CAD programs require this, others provide this as an option.

Step 6: Set snaps and drawing modes. *Snaps* allow you to draw precisely to geometric features of objects, such as the center of a circle or perpendicular to a line.

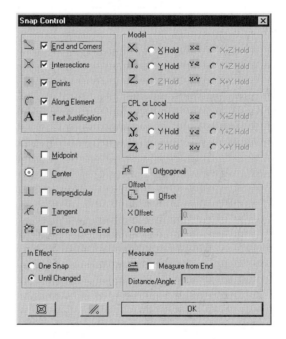

The dialog box in CADMAX for selecting snaps.

Drawing modes include visual cues, such as the grid and cursor style, and drawing aids, such as snap modes, and orthographic or isometric modes.

Step 7: Load text fonts and define text styles. Many CAD packages have just one or a limited set of fonts preloaded in their drawings. To use any other font, you need to load it into the drawing; in some cases, you may need to convert the font file into a format understood by the CAD package.

Once the fonts are loaded, create the text styles that define how the font appears in the drawing. See chapter 5, "Fonts and Patterns, Linetypes and Widths."

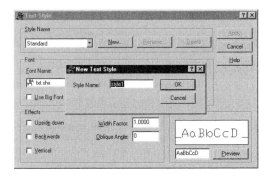

Creating a text style in AutoCAD.

Step 8: Load linetypes and set the linetype scale. Some CAD packages come with hardwired linetypes for every drawing; other CAD packages require that you load the linetypes from a file before they can be used.

Unless you have a specific reason for not doing so, I find it is faster to have every linetype loaded into the drawing, then remove the unused ones when the drawing is complete.

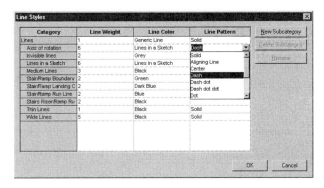

Specifying linetypes in Revit.

Standard Drawings and Templates

In some cases, you may need to define your own linetypes, if they are not provided with the CAD package. The linetype scale usually applies globally to all linetypes in the drawing, but some CAD packages allow you to define the linetype scale on a per-object basis. See Chapter 5.

Step 9: Create layers and assign colors. If the CAD package does not provide a set of standard layer names that you can use, this may well be the most tedious part of the template creation process. In the worst-case scenario, you'll need to type the name of every layer. Some CAD programs allow you to import a set of layer names from a file or other drawings. When creating each layer, you'll need to give it:

- A name
- Default color
- Default visibility
- Default linetype
- Default line width
- Plotstyle
- Plot-noplot status

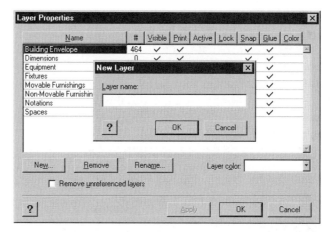

Creating new layers in Visio.

Step 10: Load or create hatch patterns. Just like with linetypes and fonts, your CAD package may require that you load the hatch pattern from a file before it can be used in the drawing.

If the CAD package does not include the patterns your discipline needs, you may need to create the pattern yourself — or obtain the pattern from another source.

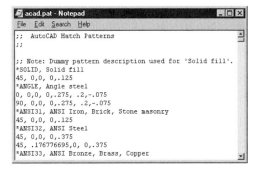

To create a custom hatch pattern in AutoCAD, you must edit the Acad.Pat file.

Step 11: Load or create references to symbol libraries. The most common "symbol" is the drawing border and title block. Many CAD systems include predrawn borders for several drawing sizes and international standards. Ideally, these elements are not inserted in the drawing, but are referenced (xref'ed) so that they can be updated automatically, should the need arise. As well, the title and border are usually placed in layout mode (a.k.a. paper space). Finally, your CAD system may need to know where the symbol library is located. See chapter 2, "Naming Drawings and Creating Symbols."

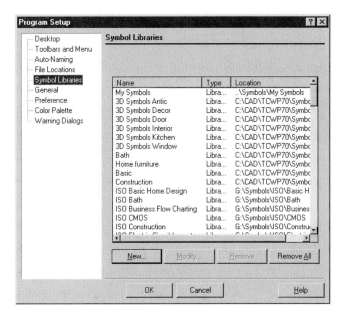

Specifying the location of symbol libraries in TurboCAD.

> **TIP GRAPHIC SCALE**
> Add a graphic scale to your template drawing. It should show a line 1" long, with the caption:
>
> **One Inch on Original Sheet**
> ├─────────────┤
>
> This lets the checker know whether the drawing was plotted full-size, half-size, or enlarged or shrunk by a photocopier to an arbitrary size. This graphic scale should always be shown on scale drawings.
>
> You may wish to include a second graphic scale of the type usually shown on maps, showing distances in the model frame of reference.

Step 12: Set dimension variables. If the CAD package does not preset dimension variables (sometimes called "dimvars") for the standard you follow, you will have to tweak the dimvars on your own. See chapter 7, "Standard Drawings and Templates."

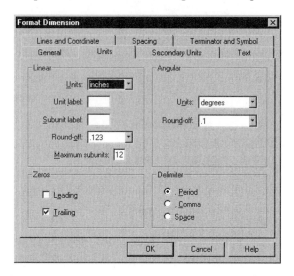

Specifying a dimension style in SmartSketch.

Step 13: Select default plotter and set plot styles. Several CAD packages provide you with fine control over the plot. For most, you'll need to specify the default plotter, especially if it differs from the default Windows system printer. You'll also need to set up the default paper size, plotting parameters, and whether batch plotting is involved.

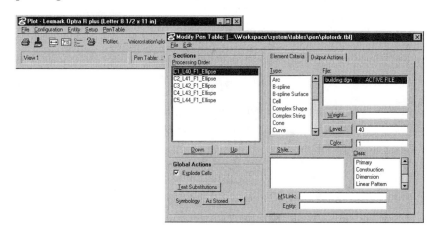

Setting up a pen table in MicroStation.

In some cases, you may not be plotting to paper, but to the Web in an Internet file format, such as JPEG or DWF.

Step 14: Create layouts. Layouts (a.k.a "paper space") define how views of the drawing are arranged on the page. If you want, you can create one layout for every type of media your office employs: one layout for vertical A-size paper, another for B-size media, etc.

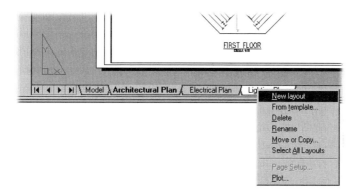

Creating new layouts in AutoCAD.

Step 15: Save the template drawing. Until now, you have probably been working with a regular drawing. To save the drawing as a template, you typically have to use the CAD software's **Save As** command, and select the template file format.

Store the template drawing in a folder specific to templates; some CAD packages automatically set up a **Template** folder.

CAD System	Template File Extension	Templates Folder
Actrix	AXT	Solutions
AutoCAD	DWT	Templates
CADMAX	CXS	Starters
DataCAD	TPL	Tpl
IronCAD	ICD	Template
MicroStation	DGN	Seed
Revit	RFT	Template
SmartSketch	IGR	Template
TurboCAD	TCT	Template
Visio	VST	Solutions

Standard Drawings and Templates

Step 16: Set permissions, send notifications, and write documentation. Once the template drawing has been created, you need to decide who gets access to it, tell people about it, and document its contents. In terms of access, you need to decide: (1) who should have access to the file; and (2) what level of access they are permitted. This is called *permissions;* you set up your firm's computers to provide users with full access, limited access, or no access at all, as follows:

- **Read-write:** Users can read (open) the drawing, make changes, and write (save) the changes. This is known as *full* access. Only the CAD manager has full access to template drawings.
- **Read-only:** Users can read the template drawing, but they cannot save changes. This is known as *limited* access. Specific users are restricted to this limited form of access.
- **No access:** Users cannot open the drawing; known as *no* access.

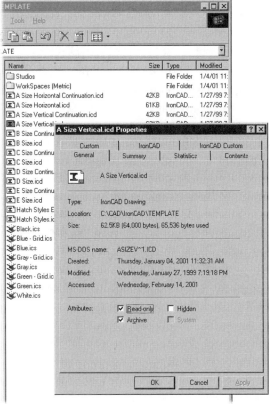

To set permissions from Windows, select the file(s), then right-click, and select **Properties**.

With permissions established, you need to let people know that the template files exist. This is usually done by sending an e-mail or posting the notice at your company's Web site. You may want to notify employees at your firm, as well as your clients and customers. At the time of this writing, some CAD vendors were starting to put an automatic notification feature into their project management software.

Finally, you need to document the contents of the template file. List the names and, if applicable, a picture of every element in the template, such as layer names, associated colors, and hatch pattern samples. It's a good idea to have the documentation in two forms: (1) print on paper and stored in a three-ring binder; and (2) posted to your Web site in HTML or PDF format.

Step 17: Make backups. Create two back-up copies of the template drawing and its documentation. Store one copy on-site (in a fire-resistant safe); store the other copy off-site (in case your office burns down). This ensures your valuable time isn't wasted by having to recreate a destroyed drawing.

After Templates are Created

Once you create template drawings, they are not cast in stone. You can at anytime edit its contents to correct errors, update standards, and add more elements. Additional elements you may wish to consider include:

· System variables
· Named views
· Standard notes
· Custom menus and toolbars
· Pre-programmed macros and programming routines.
· Third-party applications.

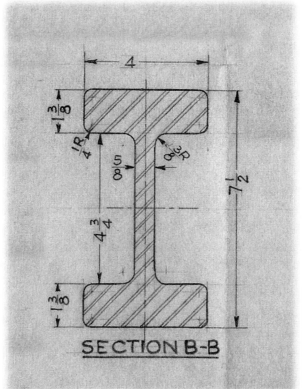
Hand drafted by Herbert Grabowski.

Writing Your CAD Standards Manual

8

In this chapter, we create the CAD standard reference book. There are two formats you should consider: paper and electronic.

One option is to document your CAD standards on paper in a three-ring binder. This lets the CAD operators quickly scan through the standards; the format can also be given to clients and contractors. The drawback to a paper-based manual is that it can be slower to search for a specific topic, and more expensive to update, than electronic documentation.

The second option is to create online documentation in HTML format (hypertext markup language, the same format used to create Web pages) or as a PDF file (portable document format, the format displayed by Adobe Acrobat). The advantage to electronic documentation is that is can be created and distributed more cheaply and quickly than paper documents. The drawback is that a Web browser can search only the currently displayed page, and it does not print very efficiently (each hyperlinked page must be printed separately). One solution is place the entire standard into one enormous HTML (or PDF) file, with a hyperlinked table of contents at the top of the document.

CHAPTER SUMMARY
This chapter concludes the work begun in the previous chapter. This chapter provides tips on how to write up the standards in a reference book or on-line documentation.

RESOURCES

The CalTrans CADD manual is an example of an HTML document that can be found at: www.dot.ca.gov/hq/esc/Engineering_Technology/Development Branch/CADD_manual/revtitle.html

The Free Software Foundation has developed standard contract language to handle the distribution of documentation at www.fsf.org/copyleft/fdl.html

And a discussion of alternative licensing schemes can be found at www.fsf.org/philosophy/license-list.html#Documentation Licenses

---

This chapter shows you several tables of contents to help you write the CAD standard. You may need to create several manuals:

Generic CAD Manual (Metric and Imperial versions): The information contained in this manual applies to all projects, except as noted in the project-specific manual.

Project-Specific CADD Manual: This manual is customized for the needs of the individual project; a shorter document because it does not include the material in the generic manual.

Design Software-Specific CAD Manual: This manual contains information specific to disciplines, such as creating designs with InRoads or Architectural Desktop.

Your Client's CAD Manual: Your clients may require that your work be performed according to their standards.

The office CAD standards manual typically consists of the following parts:

· Title page
· Disclaimer page
· Table of Contents
· Purpose
· Creating a Drawing
· Plotting the Drawing
· Drawing Management
· CAD Procedures
· Index

Remember to keep the documentation up to date!

Title Page

The title page identifies the documentation and should include:

· Company or agency name and address.
· Title, such as *Standard for Electronic Documents*, *CAD Procedures Manual*, *Consultant Guidelines for CAD Drawings*, or *CAD User Manual of Instruction*.
· Revision number.
· Most-recent revision date.

Every page of the documentation should have the revision number as a header or footer.

Disclaimer Page

The disclaimer page follows the title page and announces to readers and users that your firm is not responsible for problems arising from use of the standard.

You may also want to prohibit the sale and modification of the standards. Since CAD standards are still emerging, it pays to share your standards with other firms. This helps improve the standardization of the CAD industry and its drawings.

Table of Contents

The table of contents forms the outline of the entire documentation. Here is a guide you can use for your standards book:

1. Purpose
  a. Where to Go for Help
  b. Sources of Documentation
  c. Monthly User Group Meeting

2. Creating a Drawing
  a. Filenames
      i. Drawing Log
      ii. Drawing Management
  b. Sheet Numbers
      i. Title Blocks
      ii. Sheet Borders
  c. Units
      i. Dimensions
      ii. Elevations and Datum
  d. Coordinates
      i. Grid Lines
      ii. Match Lines
      iii. Reference Marks
  e. Use of Symbols
      i. Reference Tags
      ii. Standard Office Library
  f. Symbol Naming
  g. External References
  h. Layers
      i. Names
      ii. Colors
      iii. Linetypes
      iv. Standard Office Library

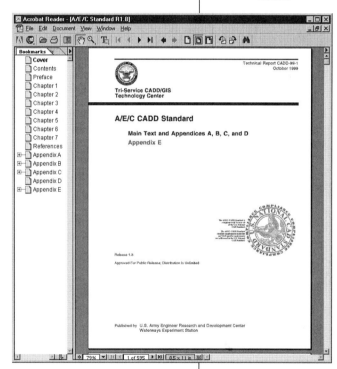

An example of a CAD standards manual distributed as a PDF document. This manual was created by Tri-Service CADD/ GIS Technology Center.

- i. Hatch Patterns
    - i. Scale Factors
    - ii. Standard Office Library
- j. Line Widths
- k. Text
    - i. Fonts
    - ii. Notes

3. Plotting the Drawing
   a. Plot Scales
   b. Sheet Sizes
   c. Ink Jet Plotter Use
   d. Laser Printer Use
   e. Batch Plotting

4. Drawing Management
   a. Backups
   b. Archives
   c. Accessing the Network
      i. Network Schematic
      ii. Network Security
      iii. Anti-virus
   d. The Menu Systems
      i. Screen Menu
      ii. Icon Menu
      iii. Tablet Menu
   e. File Extensions
   f. Translating Drawings
      i. Via DXF
      ii. Via IGES
      iii. Via DWG

5. CAD Procedures
   a. Flow Diagram for Project Drawings
   b. Adding to the Symbol Library
   c. Adding Hatch Patterns and Linetypes
   d. Function Key Definitions

It is better to be terse and symbolic, rather than long-winded in the text. Remember that this is a reference that will be scanned; no one will sit down to read it from cover to cover.

Here is the table of contents from the Tri-Service CADD/GIS Technology Center's *A/E/C CADD Standard* (Technical Report CADD-99-1). From the use of words like "design cube" and "seed files," you can tell that the emphasis is on MicroStation.

1. **Introduction**
    Acronyms
    Scope
    Purpose
    Background
    International System of Units (SI) Considerations
    Future Technologies
    Interchangeable Terminology
    Target Systems
    Additions/Revisions
2. **Drawing File Organization**
    Design Cube
        Available drawing area
        File accuracy (units)
        Drawing units/working units recommendations
        Origin (Global Origin)
    Model Files and Sheet Files
    Electronic Drawing File Naming Conventions
        Industry Standard model file naming convention
        Industry Standard sheet file naming convention
        Tri-Service Optional model file naming convention
        Tri-Service Optional sheet file naming convention
    Coordination Between Sheet File Name and Sheet Identifier
3. **Graphic Concepts**
    Presentation Graphics
        Line widths
        Linetypes/styles
        Line color
        Screening (halftoning)
        Text styles/fonts
        Plotting
    Border Sheets
        Sheet sizes
        Title block
    Drawing Scales

Dimensioning in Metric (SI)
    Millimeters
    Meters
    Large units of measure
    Dual units

4. Level/Layer Assignments
  Levels/Layers
    Level/layer naming conventions
    ISO format
  Model Files
    Level/layer assignment tables
    Border sheets
    Seed files/prototype drawings
    Reference files (XREFs)
  Sheet Files
    Level/layer assignment tables
    Development of sheet files

5. Standard Symbology
  Introduction
  Electronic Version of the Symbology/Elements
    Deliverables
    Line styles
  Tabulated Version of the Symbology/Elements
  GIS-Related Symbols

6. Tri-Service A/E/C CADD Standard Implementation Tools

7. Deliverables and Data Exchange
  Delivery Media
  Format
  Documentation
    Hard copy
    Ownership

8. References

Appendices
A. Model File Level/Layer Assignment Tables
B. Sheet File Level/Layer Assignment Tables
C. Simplified Model File Level/Layer Assignment Tables
D. Color Comparison and Associated Line Widths
E. A/E/C CADD Symbology

And here is the abbreviated table of contents from United States Coast Guard's Civil Engineering Technology Center *Civil Engineering Computer-Aided Design Database (CE-CADD) Version 6.01c4*. (CE-CADD is the name of an AutoCAD software add-on that assists drafters in creating drawings that meet the standard.)

Preface
1. General
2. Graphic Concepts
3. Drawing File and Set Organization
4. Layering
5. The Main CE-CADD Toolbar
6. CE-CADD Drawing Setup
7. CE-CADD Descriptors and Dimensions
8. CE-CADD Editing and Assistance Tools
9. CE-CADD Draw Functions
10. Plotting CE-CADD Drawings
11. Converting Drawings from v5.2 to 6.01
12. Deliverables

Appendices
A. Toolbar Map
B. Quick Reference
C. Sheet Sizes
D. Text Heights
E. Example Sheet Borders/Title Blocks
F. Schedules
G. Layers
H. Symbology
I. Glossary
J. References

Standard Office Library

Not only should the office standards be embedded in a prototype drawing, but the standards should be liberally illustrated throughout the documentation.

The following figure illustrates the format of pages that document visual elements, such as linetypes, the symbol library, hatch patterns, and text fonts (if more than one font is used).

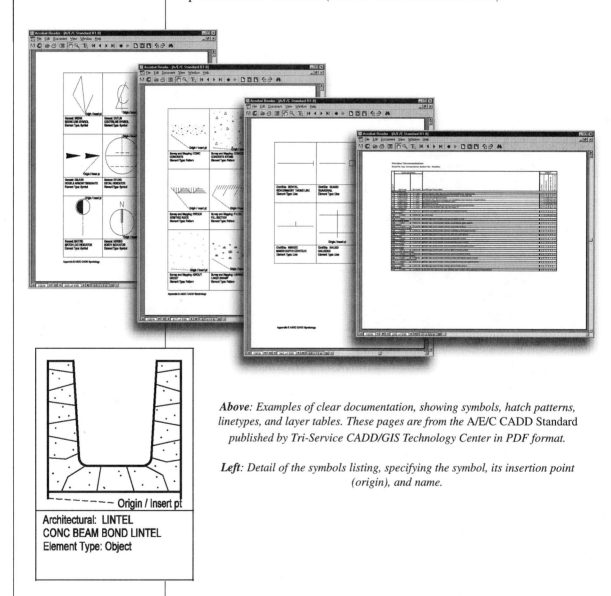

Above: Examples of clear documentation, showing symbols, hatch patterns, linetypes, and layer tables. These pages are from the A/E/C CADD Standard published by Tri-Service CADD/GIS Technology Center in PDF format.

Left: Detail of the symbols listing, specifying the symbol, its insertion point (origin), and name.

After the Document is Complete

After completing the CAD standard documentation, take a break! But you will soon find that the document is a living standard. From time to time, you will have to update it. I recommend using the word processor's automation tools to keep the document accurate and up to date, as follows:

- Use the spell checker and grammar checker to produce a well-written document. Spelling mistakes and bad grammar reduce the credibility of your office standards.
- Use the Table of Contents (TOC) and Index generators to automatically renew the TOC and index.
- Use your word processor's hyperlink feature to create links between the TOC and index and the body of the document. Save the document in the word processor's native format (eg. Word's DOC format) and save a second copy in HTML format.
- Perform editing on the DOC file; distribute the HTML file to CAD users.
- Create hyperlinks between the documentation and the CAD program.
- For low-cost distribution of the standard, post the HTML file on your company Web site.

BILL OF MATERIAL

PART No.	No. of UNITS	No. Per UNIT	DESCRIPTION	MATERIAL	STOCK REQ. No.	REMARKS
		1	PLATFORM ASSY.			
-2		5	4" @ 6.25# [, 10'-8 5/8" LG	STEEL	2503625	
-3		1	3"-3"× 1/4" L , 18'-6" LG.	"	2501103	
-4		1	—"— , 17'-10 3/8" LG	"	"	
-6		4	2 1/2" × 2 1/2" × 1/4" L , 3'-10 1/2" LG	"	2501104	
-7		7	—"— , 3'-6 1/2" LG	"	"	
-8		62 FT	2"× 2"× 1/4" L	"	2501102	
-9		27 FT	4"× 1/4" FLAT (FIT IN FIELD)	"	2500305	
-10		4	3 LB GRATING 2'-0" LWD.× 5'-6" SWD.			PEDLAR TYPE
-11		2	5/16" WELDED CHAIN, 2'-6"	"	2800103	
-12			16 GA × 2" ARMCO INTERLOCKING GRATING	"		
			9 PCS 4'-4 1/2" LG. EA.	"		
-13		33	PCS 2'-0" LG. EA.	"		
-14			2 × 2 × 3/16" LG. , 6'-3" LG.	"		
-20		1	ACCESS LADDER ASSY.			
-21		2	3"× 3/8" FLAT , 9'-0" LG	STEEL	2500506	
-22		8	7/8" Ø ROUND, 1'-4 3/8" LG	"	2501707	
-23		2	1 1/4" Ø STD. BLK. PIPE	"	3000007	
-24		2	1/2" RED HEAD ANCHOR	"	2302513	
-25		2	1/2" N.C. HEX. HD. BOLT, 1 1/2" LG	"	2390151	
-26		2	9/16" FLAT WASHER		2390710	
B-16848			STD. LADDER, STEEL DET.			
B-16844			STD. HANDRAIL, —"—			

Hand drafted by Herbert Grabowski.

Working with Paper Drawings

9

Many of the previous chapters (indeed, most books on CAD) tell you how to create drawings in CAD. But what about the drawings that already exist on paper? The toughest problem in CAD involve inputting old paper drawings into the CAD package.

Thanks to fast plotters, CAD software easily generates paper drawings. The reverse, however, is much more difficult. It is difficult to change a paper drawing into a digital equivalent. The primary method is to employ a scanner to read the paper drawing, and save the scan as a raster file — essentially, taking a digital "photograph" of the lines on the paper.

CHAPTER SUMMARY

This chapter describes strategies for dealing with paper, and provides a rationale for not dealing with paper. If your firm deals with drawings on paper, then this chapter describes some of the problems and solutions to dealing with them.

Creating the raster file is the easy first step; the difficult second step is to convert the raster file into a vector drawing understood by CAD software. This chapter describes the variety of techniques invented over the last 20 years to solve the problem of raster-to-vector conversion.

A project consists of more than just CAD drawings; there are contract documents, photographs, and marketing material as well. A scanner manufacturer[*] created the following list of documents that their scanners have handled:

- Architectural drawings
- Engineering schematics
- Aperture cards
- Manufacturing diagrams
- 35 mm film
- Slides
- Transparencies
- Big posters
- Original art work
- Story boards on up to 1/2" thick dry mount
- File cabinets stuffed with many year's of paperwork
- Huge GIS quad maps
  - Street maps
  - Land plats
  - Soil and mineral deposit maps
  - Top secret aerial reconnaissance photos
  - Oversized historical blue prints with frayed edges

[*] Ideal Scanners and Systems, Inc.

Why should you need to worry about these documents? You may need to reuse existing site and facility drawings for new projects and renovations.

More seriously, the media is slowly disintegrating, particularly drawings drafted on acidic paper. Drawings created in the 1950s or earlier may have crumbled into flakes of paper by now. Even today's inkjet plotters do not create long-lived plots. (As this book was being written, however, Epson announced it had created pigment-based inks it claims lasts 125 years.)

Knowing why you should worry about paper-based drawings helps lead you to the appropriate solutions: archiving and conversion.

TIP THRESHOLDING GRAYSCALE SCANS

Eight-bit grayscale drawings represent shades of gray by an integer between 1 and 255, with 1 being pure black, 125 being medium gray, and 255 being pure white. There is no loss of detail when shades between 1 and 125 are changed to 1 (pure black), and shades between 126 and 255 are changed to 255 (pure white). This has the added benefit of reducing the size of a file by a substantial amount.

ARCHIVING DRAWINGS

If your worry is over flaking paper, then you are concerned about maintaining the drawings as records. To preserve your firm's investment in drawings, the solution is to scan the drawings and store the data digitally on media such as CD-R discs.

Archiving is easy, but dull work. One electrical power authority estimated it would take until the year 2020 for a single person to scan all their drawings into digital format.

The advantages of digital archiving are that it requires less storage space than paper drawings; plastic media is more stable than paper; and access to drawings is computerized and hence faster.

But there are disadvantages to going digital. The drawings, even though digital, cannot be used in CAD; the true media lifetime of plastic not known (estimated to be anywhere from under 10 to over 100 years).

The technology for this process is readily available and reasonably priced. Here's what you will need: scanner, software, storage, and (optionally) a printer.

> **TIP OPTICAL VERSUS SOFTWARE RESOLUTION**
>
> Some scanners boast of having very high resolution, such as 9600 dpi. Most likely this is the scanner's *software* or *scaled* resolution. The scanner interpolates to create the effect of higher resolution. The true resolution shows up on the spec sheet as the optical or hardware resolution.
>
> For example, the spec sheet of a scanner reads:
>
> **Resolution:**
> Optical resolution is 400 dpi.
> Scaled (interpolated) resolution from 100 dpi to 800 dpi.

The Scanner

You will need one or more scanners that handle the size of drawings you have in storage. Scanners range in capacity from handling tiny aperture cards to monster drawings over 4 feet wide. Different models scan in 16.7 million colors, 256 shades of gray, or monochrome. Resolution can range from as low as 25 dpi to as high as 4000 dpi (dots per inch). And they can handle media up to 54" wide, 0.6" thick, and unlimited length.

Large format scanners read a paper drawing in 8 to 60 seconds, depending on the speed of the scanner, the resolution, color depth, and the size of the drawing. Speed is expressed as ips (inches per second) at a specific resolution, such as 6 ips @ 200 dpi and 3 ips @ 400dpi. The higher the resolution, the slower the scanning speed.

Scanners include software that adjusts the scanned image for minor imperfections such as *skew* (the drawing was not fed through at a precise rightangle), *adaptive thresh holding* (automatically creating the highest contrast between light and dark areas), and cleaning up *detritus* (imperfections in the drawing).

Working with Paper Drawings   **141**

Drawings scanned at 300 dpi (even on a low-quality scanner) are more readable when scanned in at 8-bit gray than when scanned at 1-bit (monochrome). There are fewer "jaggies" on diagonal lines, and better legibility of fine details.

The output from the scanner is to either direct a printer or it can be stored in a variety of common raster and vector file formats. A high-speed SCSI or USB interface is used to handle the huge amount of data generated by a large-format scan.

Top-of-the-line scanners are $25,000 (US funds) new, but the cost can be reduced to under $10,000 by purchasing a less powerful model, or a reconditioned unit.

Although large format scanners are required for D- and E-size drawings, don't overlook 11" x 17" scanners with document feeders. These are useful for scanning contract documents, and can scan at speeds ranging from 15 to 50 pages per minute.

The Software

Although the scanner includes some capability to correct many scanning problems, you typically purchase additional software to clean up, print, archive, and retrieve the scanned drawings. The software performs automatic and manual functions. Automatic means the software tries as best as it can to do the work itself; manual means the CAD operator performs the work.

- **Automatic cleanup** removes speckles, small black dots that shouldn't be part of the drawing (a.k.a. noise). Speckles that the software proposes to erase are displayed in a second color.
- **Automatic deskew** straightens out the image because the drawings usually are not scanned straight.
- **Manual deskew** means the CAD operator does it himself.
- **Intelligent picking** selects raster data that can be easily recognized, such as lines, arrowheads, and arcs/circles.
- **OCR** (optical character recognition) converts the appropriate portion of the raster images into text; this usually includes a spell checker to correct words that were poorly converted.

The raster editing software is available in three styles: stand-alone, batch mode, and CAD environment. Stand-alone software reads a raster file, you edit it, then save it. In batch mode, you set the cleanup parameters; the software then reads all raster files in a subdirectory (folder), processes all of them one at a time, and saves the result. In the CAD environment, the software operates inside

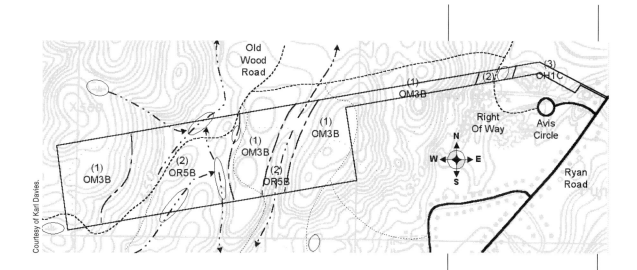

This park map of Sawmill Hill was drawn in Visio on top of a scanned topographic map.

the CAD program, allowing you to mix raster images and vector drawings. (Most CAD programs can load and display raster images, but don't provide any raster editing tools useful for correcting scanned images.)

While the scanner might be able to process drawings of unlimited length, the software isn't as omniscient because of limitation in the computer system (namely, storage space and maximum addressable memory). One package, for example, is limited to a half-billion pixels, which is the equivalent of a 4' x 15' drawing scanned at 200 dpi in monochrome. Remember: the higher the resolution, the smaller the drawing size that can be scanned.

Other useful software you might consider: Archival software that stores the image files in compressed format on the backup device. Retrieval software lets you select, view, and print archived drawings. This software can automatically create a database entry based on the text it reads in the title block area.

Sizable Storage

A backup device is required that stores the compressed image files on the archival media. The choice of media (digital tape cartridge, cartridge drives, CD-R and DVD-R discs, or optomagnetic disks) depends on your budget and capacity requirements.

If you need to regularly access the archived drawings, your firm may need to consider a jukebox retrieval system where a robotics device finds and selects the tape or disc and inserts it into the reader.

As this chapter was being written, Quantum's DLTtape Group introduced its Super DLT tape drive ($6,000) that reads and writes 110 GB tape cartridges ($149) at 11 megabits per second.

The Printer

A raster output device, such as B-size laser printer or E-size inkjet plotter, makes hardcopies of the archived drawings. Some scanners allow the printer to be attached directly, creating an enormous copying machine: input paper drawing in the scanner, and output copied drawing from the printer.

RASTER ISSUES AND CALCULATIONS

Raster files are anathema to CAD operators. Raster files are stupid, large, and unwieldy. On the other hand, raster files are well-documented and completely predictable — unlike CAD vector files.

Dots Per Inch

The primary metric by which raster files are measured is by their resolution. This is usually defined by dpi (dots per inch). In metric countries, it is dots per centimeter. The word "dot" actually refers to a pixel. There is a direct relationship between dpi, color depth, and storage requirements.

You can calculate the amount of disk space an uncompressed image takes up: multiply the size of the image by its resolution (expressed as dots per inch).

For example, scanning an A-size ($8\frac{1}{2}$" x 11") document at 200 dpi results in a 456 KB monochrome file. Here's the math:

Media area = Height x width
 = 8.5" x 11" (A-size)
 = 93.5 square inches

Resolution = 200 dpi x 200 dpi
 = 40,000 dots per square inch

File size = Media area x Resolution / 8 bits per byte
 = 93.5 x 40,000 / 8
 = 467,500 bytes / 1024 bytes per kilobyte
 = 456 KB

The A-size drawing scanned in monochrome at 200 dpi creates a half-megabyte file, uncompressed.

Color Depth

The previous calculation was for a monochrome scan. The raster file needs just one bit to store a dot in monochrome (a.k.a. 1-bit color). That bit is either white or black. Adding grayscale or color increases the amount of storage space. Grayscale images typically display 256 shades of gray, while color images often (but not always) employ 16.7 million or more colors.

Rather than simple multiplication, color depth involves factors. For example, the reason "true color" images have 16.7 million colors is because the computer uses 24 bits to define the color (8 bits each for red, green, and blue.)

Bits	Colors or Shades of Gray	Common Name
1	2	Monochrome or bitonal
8	256	Grayscale
16	65,000	High color
24	16.7 million	True color

Multiply the monochrome file size (456 KB) by the number of bits in the table above. (The number of bits is more common called the "color depth.")

For example, the size of a grayscale scan is:

File size = Monochrome file size x Color depth
 = 456 KB x 8 bits
 = 3,648 KB / 1024 KB per megabyte
 = 3.56 MB

To extend the example further, the same A-size drawing scanned in full color (24 bits) creates a 11 MB file, uncompressed.

Compression Issues

The effect of compression on reducing the file size cannot, unfortunately, be predicted. The reduction depends on the efficiency of the compression algorithm and the complexity of the image.

Some compression algorithms work better with grayscale images, while others work better with color images. Some compress algorithms are better suited to photographic images, while others work better with linework.

> **TIP MORE THAN 24 BITS**
>
> You may hear of scanners and graphics boards capable of 32 bits of color depth.
>
> When applied to scanners, the scanner uses the extra bits of color to generate, in its opinion, a better image. Only 24 bits of color are delivered to your computer.
>
> When applied to graphics boards, the extra 8 bits are used for transparency to create the see-through look common in games and Windows XP.

The Effect of Resolution On Scan Quality

The higher the resolution, the better the scan. But higher resolution also leads to longer scan times and larger file sizes, which are cumbersome to deal with. Larger files take more storage space, take longer to load into software, and longer to process. Larger file sizes can be partially overcome with file compression. File compression, however, can increase the time it takes to load and save raster files.

The images on these two pages show the effect of increasing the scan resolution. The entire drawing is shown on this page; a detail from the center of the drawing is shown on the next page at three resolutions: 50 dpi, 200 dpi, and 1200 dpi (the highest hardware resolution for the scanner I used).

The figure on this page is a scan of a drawing created with pencil on paper. The center portion (shown by the rectangle) has been enlarged on the facing page.

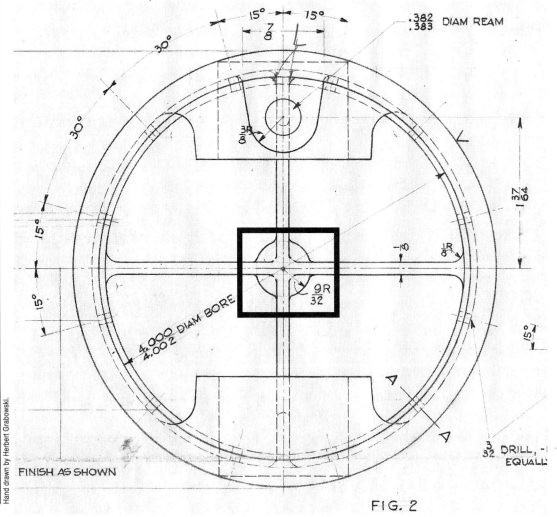

50 dpi

200 dpi

1200 dpi

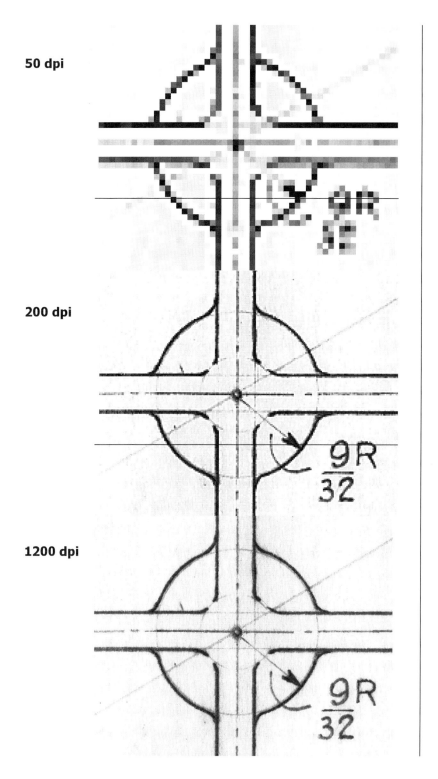

The three figures on this page are an enlargement of the center of the scanned image on the facing page. Each enlargement shows the improvement from of scanning the paper drawing at three different resolutions.

Working with Paper Drawings

JBIG2 AND JPEG 2000

Two new raster formats may become important in the future:

JBIG2 compresses images optimized for monochrome documents, resulting in files three to five times smaller than current fax standards. JBIG2 was developed by the ITU and ISO/IEC standardization bodies. The new format supports multi-page document compression, which increases compression by another factor of two due to its reuse of character shapes. Images are decompressed at speeds of over one gigapixel per second.
pr.jpeg.org/press.htm

JPEG 2000 has a higher compression efficiency, and supports multiple resolutions in a single file. Compression ranges from lossless to lossy.
www.jpeg2000info.com

Where is GIF?

Another very popular raster format is GIF (graphic interchange format) invented by CompuServe in the mid-1980s. GIF is good at producing small files without losing details or adding artifacts around text and lines (as does JPEG). Its popularity declined in the late 1990s when Unisys, the owner of a patented compression algorithm used by GIF, began demanding royalties. Rather than pay up, some companies, such as Autodesk, simply dropped support for GIF. At the time of writing this book, Unisys was demanding royalty payments from some large Web sites displaying GIF images.

Common alternatives to the problematic JPEG and GIF raster formats include TIFF, PNG (portable network graphics), and RLE (run length encoded).

Why You Shouldn't Use JPEG

The JPEG file format is a very popular format for large, photographic images such as those created by digital cameras. (JPEG is short for joint photographic expert group). The format is popular because it is able to store an image in a file 20 to 30 times smaller than other formats.

For example, here are the statistics for type different types of images saved in three formats: uncompressed (the full size of the image), TIFF (the most popular format in desktop publishing), and JPEG:

Image	Engineering Drawing (Grayscale)	Digital Photograph (Color)
Common Resolution	1200dpi	72 dpi
Typical Size	7,500 x 6,500 pixels	1600x1200 pixels
Color depth	8 bits	24 bits
Uncompressed *	48MB	5.6MB
TIFF LZW	24.8MB	3.0MB
JPEG 50%	1.9MB	0.17MB

* BMP or uncompressed TIFF

While JPEG is able to greatly compress large image files, it has a fundamental flaw: the method by which it performs compression creates artifacts. These are unwanted additions to the image. In a digital photograph, they are barely visible; artifacts become much more noticeable in line drawings. For that reason, never use the JPEG file format to store drawings.

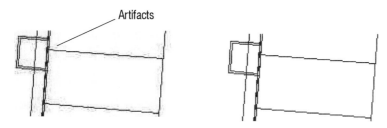

A highly compressed JPEG image with artifacts (left), and a clean TIFF image (right).

CONVERTING DRAWINGS

In the mid-1980s, programmers and their marketers were sure that CAD's most perplexing problem was close to being solved: feature recognition. Their trumpeting misled many users (even to this day) to think that it is possible for software to convert a scanned drawing image into an accurate vector CAD file. Let's look at the problem more closely to understand why it is so difficult to solve.

When you look at a drawing, your eye and brain make out the lines, circles, and text as independent yet connected objects. You are able to read text stained with coffee; you understand lines with arrowheads to be dimensions and leaders; you know the difference between the lines making up the drawing border and the lines representing the building walls. Early in the twenty-first century, computers still cannot differentiate very well.

The reason the computer does a poor job of feature recognition is that currently the only technology that "reads" a paper drawing is the raster scanner. Some work did go into developing a vector scanner, which was limited to tracing contour maps.

The raster scanner contains a row of light sensors, typically 400 sensors per inch of scanner width, which works out to about 20,000 sensors for an E-size scanner. The light sensor is often a CCD (charge-coupled device), the same lens used in video and digital cameras.

As the drawing passes through the scanner, the CCD takes a reading every split second — 400 readings per inch of drawing, from which comes the 400 dpi resolution rating. With each reading, each sensor reports whether it detects light (the paper's background) or dark (a drawn line). Most scanners can also sense up to 256 relative levels of light and dark, and these are called "levels of gray."

The result of the data collection is a matrix of values, dark and light. Unfortunately, reading the drawing in 1/400" increments loses the vector information.

Now it is up to the computer software, called "raster to vector converters" to make sense of the dots. The software that exists today can determine lines and arcs. Some software recognizes text after a fashion by isolating all objects smaller than 40 pixels square. High-end raster-to-vector conversion software claims to be able to reconcile dimensions.

Finally, the conversion of paper drawings into electronic format simply cannot recreate layers, attributes, colors, database links and even has difficulty with linetypes. In short, all the data that you expect to find in an electronic drawing cannot be recreated — nor will they ever be — through the raster-to vector conversion process.

In the figure on the following page, the drawing of an electrical symbol has been scanned in. Note how seemingly straight lines are recorded as jagged and broken segments. The scanned image was automatically converted into a vector format, and read into CAD software.

Original Drawings Are Not Accurate

Which leads to the other problem of inputting paper drawings into CAD. We have come to expect that drawings created with CAD are accurate to six or more decimal places; we tend to forget that drawings made with a pencil are inherently inaccurate. The accuracy of a handmade drawing is approximately one width of the pencil lead or pen tip — about one decimal place of accuracy.

If you've been a manual drafter, then you know that a certain amount of fudging goes on. Lines look straight, but probably run at a slight angle to match up. Corners look square, but probably are off by a degree. Circles appear to have an accurate radius, but are probably under- or oversized. Dimensions might not measure up to their stated distance. And fillets tend to not quite join at the corner.

On top of the inherent inaccuracy of manual drafting, it was common in pre-CAD days for the drafter to declare in the scale portion of the drawing's title box: "Not to Scale." The drawing

TIP **DIFFERENCE BETWEEN VECTOR AND RASTER**

Vector drawings are defined mathematically. For example, the arc in the symbol shown here is defined by its start coordinates, its radius, and its angle. When enlarged, the vector drawing maintains its resolution.

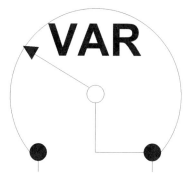

*An enlargement of a symbol drawn
as a vector drawing
shows no loss of resolution.*

Raster images are defined by pixels. In a monochrome image, each pixel is either black or white. In a grayscale image, each pixel is a shade of gray. In a color image, each pixel is a hue of color.

When enlarged, the raster image loses its resolution. This can be partially overcome through anti-aliasing, which adds extra gray pixels to make curves and diagonal lines look smoother.

*This enlargement of a symbol
converted to a raster image shows
loss of resolution. The individual
pixels are pronounced.*

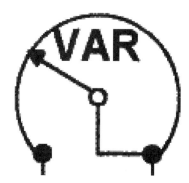

*This enlargement of the raster
image shows a slight improvement
due to anti-aliasing (the addition of
gray pixels).*

Working with Paper Drawings

looked good, but the drafter had made no attempt to accurately scale the parts.

Thus, there is no point in having really accurate raster-to-vector conversion software since the original paper drawings are themselves inaccurate.

TWO EDITING SOLUTIONS

There is difficulty in inputting paper drawings into a CAD system. The problems are inherent in every step of automatic conversion:

Step 1. The original paper drawings are inaccurate due to manual drafting practices.
Step 2. The scanner reads paper drawings as an array of meaningless dots.
Step 3. The raster-to-vector conversion software finds rudimentary patterns among the dots.
Step 4. The vectorized drawing requires extensive post-conversion editing within the CAD system.

With these drawbacks to automatic conversion, what are the alternatives? Here are four ideas:

- You may want to use automatic conversion if the accuracy and intelligence of the drawing are not an issue in your discipline.
- Another use for automatic conversion is for small jobs such as client logos and parts from vendor catalogues.
- Use text recognition software to convert raster text into ASCII format.
- The primary alternative to automatic conversion is manual digitizing. There are two methods, digitizing board and heads-up digitizing. With the digitizing board method, the paper drawing is taped to an E-size digitizing board, and drafter traces over the drawing, applying their own intelligence to feature recognition. With the heads-up digitizing method, the paper drawing is scanned and displayed as a raster image on the screen. The drafter traces over the raster image.

Some firms employ speech input to help speed the drafting process; this lets the drafter speak commands, rather than type them in at the keyboard.

The Partial Conversion Strategy

When it was recognized in the late 1980s that software was incapable of converting raster scans into perfect vector drawings, Robert Godgart came up with a radical reengineering of the process: if raster images cannot be accurately converted into vector format, then don't convert them.

He rationalized that most firms don't actually want to convert an entire paper drawing into vector. Instead, most firms use the old drawings for renovation work—perhaps changing the size of a room or rerouting the path of a highway. His re-engineering went as follows:

1. Keep the raster image in raster format.
2. Display the raster drawing as a background image within the CAD system.
3. Erase the portions of the raster drawing that no longer apply.
4. Draw the new and modified parts with the CAD package's vector commands.
5. Convert the vector lines into raster format (a trivial exercise).
6. Print out the revised drawing on a laser printer or inkjet plotter.

The approach to take when dealing with paper drawings depends on how your firm uses them. You can either ignore them, archive them in digital format, attempt to convert them wholesale to vector format, or use them as a raster background to vector-based revisions.

CASE STUDY: THE DIGITAL DRAWING

A log homebuilder in my neck of the woods implemented AutoCAD almost a decade ago. At first, it was merely a replacement for the drafting board. But the firm has been lucky enough to hire a CAD operator who had grand ideas for what CAD could do. Through his efforts, the log homebuilder eliminated paper drawings from the design construction process. To see how this is possible, let's follow the process from conceptual design through to finished product.

A customer is interested in having a log home built on a remote island site. When she comes into the builder's office, the salesman invites her to sit down at a computer with a CAD operator. The operator brings up renderings of past log home projects. When the customer sees a home that catches her fancy, the CAD operator brings up the floor plan. As the customer discusses the changes she would like with the salesman, the CAD operator enters them into the digital drawing.

With the final changes in place, the CAD package generates a 3D view from the 2D floor plans, then renders the wireframe drawing to let the customer appreciate the proposed design changes. With the customer's approval, the CAD software generates these lists:

- A price list with two prices (manufacturing cost and retail price) is printed in the showroom. The customer and salesman haggle over the final selling price.
- A log cutting list is transmitted to the log yard. In the log yard, the cutting list (in NC numerical control code) instructs the saws to cut the wood needed for the house.
- A parts list of nonlog items (windows, door knobs, etc.) is transmitted to the warehouse.

The CAD system also generates assembly drawings (on paper!) for the site. When all the parts are together, they are airlifted by helicopter to the island site where the home is assembled.

Eliminating the Paper Trail

The log-home story was just the first time I'd heard of all-digital drawings. The most common application is with lathes and other metal milling machines, which do not need drawings to be printed. Since this type of machinery is typically computerized, third-party software exists that converts a CAD drawing into the NC code that instructs the machines.

Another example is survey data. At one time the surveyor laboriously wrote notes in waterproof notebooks, but now survey instruments record data electronically in their own memory banks. At the end of the day, the data is dumped to the on-site computer or transmitted via modem back to the office. Third-party software translates the survey data into a CAD drawing.

The exciting part of all this is that no paper drawings are used in the manufacturing process, which eliminates:

- The time needed to plot, trim, and collate a set of drawings on paper.
- The errors made in reading data and 3D drawings converted into 2D conventions.
- The expense of reentering data from paper sources (notebooks, other drawings) into the computer.

Thus when you set up your CAD system, consider whether paper drawings can be eliminated entirely, perhaps except for check plots. If your clients or other departments in your firm use data in digital form, then chances are you will be able to find a way to eliminate the need for paper drawings.

SECTION "B-B"

Hand drafted by Herbert Grabowski.

Outsourcing and Extranets

10

A service bureau can be considered a type of "office overload" service. Service bureaus provide one-of services when it is too expensive to purchase the equipment for your office. A typical service bureau offers some of the following services:

- Training in CAD software and computer hardware
- Customizing the CAD system for your discipline
- Programming add-on applications to your CAD package
- Scanning drawings electronically
- Digitizing drawings by hand
- Plotting drawings
- Translating drawings from one format to another
- Setting up a document management system
- Repairing specialty hardware
- Rendering and animation from 3D CAD files
- Color slide, videotape, and hard-copy output
- Loaner equipment
- Contract drafters
- Project hosting

You may be charged by the hour (in the case of programming and customization), by the course, or by the square foot, as in plotting and digitizing).

CHAPTER SUMMARY

In this chapter, you learn about working with a service bureau, including extranets, the services they provide, and their pricing.

WHAT IS A SERVICE BUREAU?

A service bureau provides CAD services when your own firm cannot. You may find your firm's plotters aren't going to produce hard copy drawings in time for the deadline. You've been given a job that requires an E-size digitizer to trace in drawings. You need to post your project drawings on the Web for others to access. Or the project requires the next release of CAD, which you haven't upgraded to. For extra help with your CAD work, service bureaus help out in these "office overload" situations.

There are two types of service bureaus: local and Web-based. Local service bureaus are located in computer dealerships, in blueprinting houses, or are run by private consultants. Some specialize in one aspect (digitizing or plotting), while others attempt to provide all services. Before committing to a service bureau, check their references, look over their premises, and do a trial run. While most service bureaus do their best job for you, some are opened by recently unemployed CAD drafters who lack high-quality equipment to do a professional job.

If your location has no nearby service bureau, the Internet and overnight couriers make it possible to deal with service bureaus located anywhere in your country (crossing the border from another country can be tricky). For example, you send the drawing file to the service bureau by e-mail or upload it to their Web site; you receive the plotted media back by courier. Some bureaus use offshore labor, particularly for digitizing, to help keep their prices down.

Some service bureaus (known now by the trendier term, ASP) provide mechanical design and project collaboration over the Internet. Here, Autodesk Streamline improves the resolution of a wheel part as more data is streamed to the user's Web browser.

Bureau Services

Here is a rundown of the services provided by CAD bureaus:

Input
- Scanning drawings with a scanner
- Digitizing drawings by hand
- Translating drawings from one format to another
- Contract drafters
- Web-based CAD software

Project Management
- Displaying drawings and symbol libraries on a Web site
- Sharing project drawings among group members
- Allowing group communications on a per-project basis
- Archiving of CAD data on remote servers

Output
- Plotting drawings
- Rendering and animation from 3D CAD files
- Color slide, videotape, and hard copy output.

Maintenance
- Training in CAD software and computer hardware
- Customizing the CAD system for your discipline
- Programming add-on applications to your CAD package
- Setting up a document management system
- Repairing specialty hardware
- Loaner equipment

If the bureau offers training, check whether they are recognized by the CAD company. For example, Autodesk has a network of Authorized Training Centers.

Drafters can become certified through the National Association of CAD/CAM Operators, with a certificate from an authorized training center or local college, or through the AutoCAD Certification examination.

Consultants can be certified by some CAD vendors through a registered consultant program.

OUTSOURCING PROJECT MANAGEMENT

The popularity of the Internet has led to a whole host of Web-based services for CAD. You might have heard of these services under a variety of names, such as *extranets*, *ASPs*, and *hosted service*. The three terms are similar in meaning. More specifically, an extranet is a Web service external to your company, while ASP is short for application service provider. There are a host (pun intended) of services available via the Web:

· 3D model translation and viewing
· Industry directories and building code database
· Mechanical CAD collaboration
· Calendar sharing
· Bid invitation and construction document processing
· Electronic engineering data distribution
· Construction materials e-commerce
· Java thin-client CAD viewing and markup software
· Auctions and request-for-quotation service
· Salesforce automation
· Real-time streaming audio and video via a Web-controlled site camera
· Publishing files to a Web site and CD-ROM

One consultant who specializes in tracking these firms, Joel Orr, counted 156 firms at the time of writing this book. Since the industry is new, many firms are failing. Another consultant, Paul Doherty, has counted 35 firms that have failed over the few years.

An example of a Web-based service (this one for buying and selling asphalt paving and roofing products) announcing it is going out of business.

Gallery of Web-based Services

3Dshare.com, from PlanetCAD, translates and heals (fixes mistakes) 3D CAD drawings. Files are up- and downloaded to and from the site.

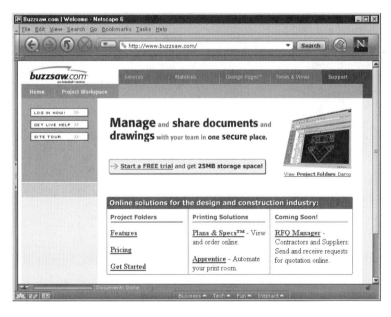

Buzzsaw.com, an Autodesk venture, provides project management services and construction bidding.

RESOURCES

For a list of extranets, see www.extranets.cc/the_list/3dshare_to_zoomon.htm.
For a list of failed extranets, see www.cyberplaces.com/404.htm.

4Specs.com provides a central location from which to search for manufacturers' specifications using CSI categories.

Alibre.com hosts a 3D mechanical CAD program that you can run through your Web browser.

BlueprintOnline.com, from ABC Imaging, lets you post drawings at their Web site for others to access.

Bricsnet.com, provides the software and services for tracking a building throughout its life — design, construction, and operation.

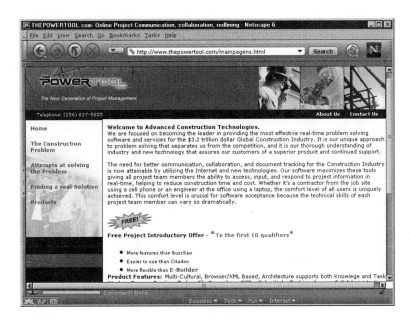

ThePowerTool.Com provides multicultural project management for the construction industry.

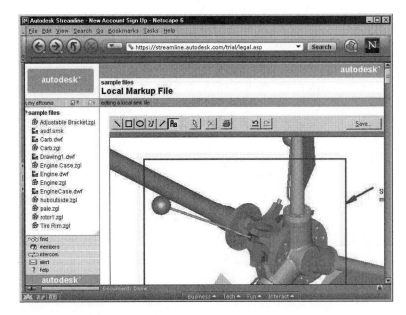

Streamline.Autodesk.com provides Web-based product management of 3D solid models generated by AutoCAD, Inventor, and Mechanical Desktop.

SERVICE BUREAU PRICING

The prices charged by service bureaus vary by bureau within cities and between cities. Here are some sample prices you can expect to pay for training:

One day	$300 to $400 per person
Three days	$700 to $900 per person

Generally, prices are higher for advanced topics such as programming, 3D Studio, and third-party applications.

For plotting, the price depends on the size of plot. Here is the range of prices charged by service bureaus:

Per hour	$20 to $60
A-size	$6 to $15
D-size	$20 to $45
E-size	$30 to $70

The range of prices is due in part to the variance in ink and media prices. Plotting on Mylar (expensive media) or a large-format rendering (large amount of ink) is more expensive than a wireframe drawing plain paper.

For digitizing, the price depends on the drawing size and the style of digitizing. The digitizing service costs more because it is more labor intensive and includes delivery of the drawing on a CD and a check plot. Here is the range of prices charged by service bureaus:

Per hour	$20 to $40	
A-size	$50 to $60	$5 for scan only
D-size	$100 to $250	$60 for scan only
E-size	$200 to $300	$85 for scan only

In the case of plotting and digitizing, avoid paying by the hour since it is difficult to determine what the service bureau accomplishes in the hour. Instead, pay by the drawing.

Outsourcing and Extranets **165**

For CAD software and project hosting, the prices range wildly. Some are free, as an incentive to try to service; others can costs thousands of dollars a month for very large projects. Some representative costs are (listed in US$):

CAD software rental	$200 per month
Project hosting	Free for 25 MB of storage space
	200 MB storage $200/month
	2.5 GB storage $625/month
	10 GB storage $1,250/month
CD-ROM archive	$25/disk (650 MB)
Web camera	$400/month (recording construction progress)

When looking for a service bureau, shop around for comparative prices. For example, perform a cost-benefit analysis before going with a hosted service. At the time of writing this book, 10 GB of storage space for a desktop computer cost $75, based on a 60 GB hard drive selling for $450. (Hard drive designers say that they typically provide double the disk space each year at the same price.) Contrast the cost of the hard drive, which you can expect to last three years, with the cost of the hosting service, which charges $45,000 to store the same amount of data for 36 months.

What other services do they provide to make their hard drive worth 600 times more valuable than a local hard drive? There are two additional services included in the cost: (1) your data is stored on secure servers; and (2) you are provided with a range of project management software.

Hosting services should give you details about their data backup procedures, and their disaster-recover plan. Is the data stored in a proprietary format useable only on the hosting services server? How do you get control over your data when:

· Your client needs electronic drawings files to be submitted at 30, 60, 90, and 100 percent completion.
· Your project ends, and you want to have the archived data in your possession.
· You decide that you want to terminate your relationship with the hosting service.
· The service goes out of business, as a number of firms did during the economic downturn period of 2000 to 2001.

While I don't recommend sacrificing quality for price, you may find at least one bureau whose prices are far higher than the average. When obtaining quotes, ask about additional charges. These might include:

- State and other taxes, such as GST in Canada and VAT overseas
- Setup charges, such as to prepare the plotter
- Material charges, such as media, ink, and videotape
- Delivery charges, such as packaging fees and COD charges

Ask about the bureau's satisfaction guarantee, including items such as delivery time, cancellation notice, refund policy, and dispute resolution.

Also inquire into the bureau's discount schedule. Most bureaus charge less per item for large amounts of work coming their way. For example, a training bureau in California gives every fourth student from the same firm free tuition.

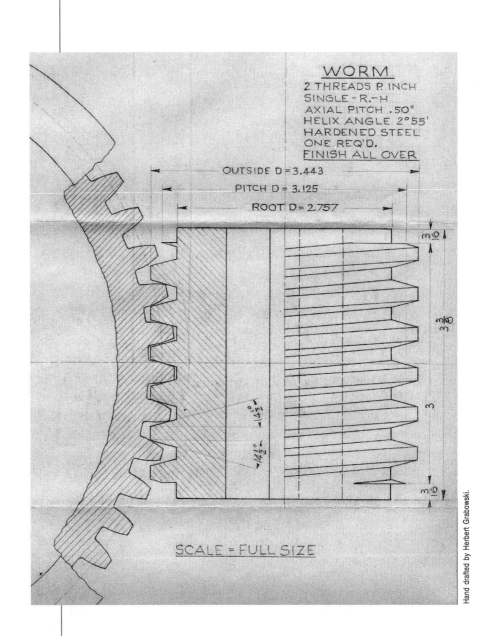

Hand drafted by Herbert Grabowski.

The DWG Format and Its Future

11

The DWG file is not documented by Autodesk. When customers ask for documentation on the format, Autodesk points them to its own DXF, the IAI's IFC, Spatial's ACIS SAT, and other documented formats that can represent drawings created by AutoCAD — though not with 100% fidelity.

There are times when only direct access to DWG will do, such as for file viewers. Several third-party developers have created toolkits for reading and writing DWG files (a.k.a. APIs). These have been popular with CAD vendors wanting to directly translate AutoCAD drawings without going through an intermediary format, such as IGES and DXF.

CHAPTER SUMMARY

This chapter describes the DWG file format produced by AutoCAD.

You also read thoughts on the future of CAD file formats, the advantages, and the problems the new formats might create.

RESOURCES

OpenDWF Alliance has documentation of the DWG file format, as well as utilities for handling DWG files: www.opendwg.org.

Autodesk has documented the DXF specification: www.autodesk.com/adsk/item/0,,140239-123112-358900,00.html.

IntelliCAD Consortium has a free CAD program that reads and writes DWG files: www.intellicad.org.

Cyco Software of the Netherlands was the very first to crack the DWG format, back in the late 1980s. They used this knowledge to write a utility to recover damaged DWG files. (This function has since been added to AutoCAD as the Recover command.) Their second product allowed you to view and print AutoCAD drawings without needing AutoCAD.

Another of the third-party developers, Sirlin, was purchased by Autodesk in the mid-1990s because, the rumor went, Autodesk itself wasn't sure what all was in the DWG file.

In the late 1990s, MarComp, was purchased by Visio Corp (now part of Microsoft) for use with its IntelliCAD and Visio Technical products' DWG translators. After buying MarComp, Visio realized that there were some unknown areas in the DWG format; to get a broad coalition of partners help it decode the format, Visio handed over the Macomp APIs to an independent organization, the OpenDWG Alliance (created and initially funded by Visio). Members of the Alliance (of which there are now hundreds) get free access to the APIs, under the condition that they report anything they figure out about DWG back to the Alliance.

Despite this massive effort at reverse-engineering, the Alliance's documentation of the DWG file format still admits there are unknown sections. In addition, ACIS data used to created 3D solid models has been encrypted, and the OpenDWG Alliance has been unable to crack the code.

Still, much of the file's format has been figured out, although the information would be primarily of interest to hardcore coders. In the following sections, I present a condensation of the document *AutoCAD R13/R14/R2000 DWG File Specification Version 2.0* provided by The OpenDWG Alliance (reprinted by permission).

THE DWG FILE SPECIFICATION

Much of the data in an R13/14/2000-format DWG file must be read at the bit level. Various parts of the drawing use data in compressed forms. Generally, the compressed forms allow for compression of common data, usually values like 0.0 and 1.0 for double-precision real numbers, 0 and 256 for short integers.

The method for interpreting the code is to read the first two bits. These indicate either the size of the data to follow, or the actual value for the common values.

All objects in DWG files since Release 13 are referred to by *object handles*. Certain object handles in AutoCAD have ownership relations with other objects. In some cases, the handle is stored as an offset from another handle, and the code indicates how the offset is to be applied.

The DWG format also uses *seekers*, which are objects that indicate either an absolute address in the file, or an offset from some known address.

Another concept used in DWG files is the *sentinel*, which consists of 16 bytes of data used for file recovery purposes.

AutoCAD DWG files use a modification of the standard *cyclic redundancy check* (CRC) as the error detecting mechanism. The modification made to CRC is that a starting value different from 0 is used. The result of the CRC is XORed with a *magic number*; this method is used extensively in pre-R13 files, but seems only to be used in the header for R13 and beyond.

General Structure

The structure of the DWG file changed between Release 13c2 and c3, but has remained consistent since then. The following notations regarding c3 indicate the differences.

The general arrangement of data is as follows:

FILE HEADER
    Version ID
    Image Seeker
    Unknown Section
    DwgCodePage
    Section-Locator Records
    CRC (cyclic redundency check)
    Ending Sentinel
DWG HEADER VARIABLES
    Beginning Sentinel
    Size of the section
    Data (system variables)
    CRC
    Ending Sentinel
CLASS DEFINITIONS
    CRC
    Sentinel
    Padding (R13c3 and later)
    Image Data (pre-R13c3)
    Object Data
    All entities, table entries, dictionary entries, etc. are in this section.
OBJECT MAP
UNKNOWN SECTION (R13C3 AND LATER)
SECOND HEADER
    Beginning Sentinel
    Data
    Ending Sentinel
IMAGE DATA (R13c3 and later)
    Start Sentinel
    BMP and/or WMF data
    End Sentinel
EXTENDED ENTITY DATA
PROXY ENTITY GRAPHICS
EOF (end of file)

File Header

The file header contains the following data:

Version ID

The first six bytes are "AC1012" for R13, "AC1014" for R14, and "AC1015" for R2000.

The next 7 bytes, starting at offset 0x06, are 6 bytes of 0 (in R14, there are 5 zeros, plus the **AcadMaintVer** system variable) and a byte of 1. The OpenDWG Alliance has occasionally seen other values here, but their meaning (and importance) is unclear.

Image Seeker

At 0x0D is a *seeker* (a 4-byte long absolute address) for the beginning sentinel of the image data.

Unknown Section

Bytes at 0x11 and 0x12 are still unknown; usually contains 0.

DwgCodePage

Bytes at 0x13 and 0x14 are a raw short indicating the value of the *code page* for this drawing file. The code page specifies the human language for the drawing, such as US English, Canadian French, or Swedish.

Section-Locator Records

At 0x15 is a long that reports how many sets of *recno/seeker/length* records follow. The records are as follows:

0   Header variables (covers beginning and ending sentinels).
1   Class section.
2   Object map.
3   *(C3 and later)* A special table with no sentinels. See "Unknown Section (R13 C3 and later)." The presence of the fourth record indicates that the C3 file format applies.
4   In Release 14 points to a location where there may be data stored. The OpenDWG Alliance has seen only the **Measurement** system variable stored here. Up to six sets have been seen in this section; the meaning of the sixth is unknown; the OpenDWG Toolkit generates files with the first five sets only.

CRC

For BOF (beginning of file) to this point. Uses 0 for the initial value.

Sentinel

A 16-byte sentinel follows the CRC.

DWG Header Variables

The header variables section indicated by section-locator 0 has the following form:

```
Beginning sentinel
Size of the section (a 4-byte long)
Data (system variables, such as PlineGen, and possibly other data at the beginning)
CRC (covers the stepper and the data)
Ending sentinel
```

Class Definitions

This section contains the defined classes for the drawing. In R14, there is a flag indicating whether objects can be moved or edited. The *itemclassid* indicates classes that produce entities, and those that produce objects.

The CRC is followed by a 16-byte sentinel.

Padding (R13c3 and Later)

This section contains 0x200 bytes of padding that can be ignored. Occasionally AutoCAD uses the first four bytes of this area to store the value of the **Measurement** variable. This padding was evidently required to allow pre-R13C3 versions of AutoCAD to read files produced by R13C3 and later.

Image Data (Pre-R13c3)

The BMP (bitmap) or WMF (windows metafile) preview image of the drawing file, if any. It is stored here only for pre-R13c3 files. Later versions place the data at the end of the file.

Object Data

This section holds the actual objects in the drawing. These can be entities, table entries, dictionary entries, and objects.

This second use of *objects* is somewhat confusing; all items stored in the DWG file are "objects," but only some of them are object objects. Others are entities, table entries, etc.

The objects in this section appear in any order. The complete list of objects in a Release 2000 DWG file are:

3Dface
AppID
AppID Table[1]
Arc
Attdef
Attrib
Block[2]
Block Control
Block Header
Circle
Dictionary[3]
DictionaryVar
Dimension Aligned
Dimension Angular 2-line
Dimension Angular 3-point
Dimension Diameter
Dimension Linear
Dimension Ordinate
Dimension Radius
Dimstyle
DimStyle Control[1]
Ellipse
Endblk
Group
Hatch
IDBuffer[4]
Image
ImageDef
ImageDefReactor
Insert
Layer
Layer Control[1]
Layer_Index
Layout
Leader
Line
Linetype Cotnrol[1]
Ltype
LwPline[5]
Minsert
Mline
MlineStyle
Mtext
OLE2Frame
Point
Polyline 2D
Polyline 3D
Polyline Mesh
Polyline Pface
Proxy
RasterVariables
Ray
Region, 3Dsolid, and Body[6]
Shape
Shape Control[1]
ShapeFile
Solid
SortEntsTable
Spatial_Filter[7]
Spatial_Index
Spline
Text
Tolerance
Trace
UCS
UCS Control[1]
Vertex 2D[8]
Vertex 3D

Vertex Mesh[9]
Vertex Pface[10]
View
View Control[1]
Viewport Entity
Viewport Entity Control[1]
Viewport Entity Header
Vport
Vport table[1]
Xline
Xrecord

Notes:

[1] Undocumented.

[2] The BLOCK_RECORD entity seems to have all the goodies that show up in a BLOCK entget, except for the common parameters. The actual BLOCK entity seems to be almost a dummy.

[3] A list of pairs of string/objhandle that constitute the dictionary entries.

[4] Holds list of references to an xref.

[5] Lightweight polyline.

[6] These are all ACIS entities. This is encrypted data that the OpenDWG Alliance has not been able to break. If this data is not present, however, generally AutoCAD will refuse to load the entity. This is why the alliance has not been able to originate ACIS in DWG.

[7] Used to clip external references.

[8] Neither elevation nor thickness are present in the 2D VERTEX data. Both should be taken from the 2D POLYLINE entity.

[9] Same as VERTEX 3D except for type code.

[10] Release 13 DWG files seem to have color and linetype data for all PFACE VERTEXs, but Release 12 and the **SaveAsR12** command seem to omit color and linetype when writing out the location vertices.

Objects have the following general format:

- MS : Size of object, not including the CRC.
- BS : Object type.

2000 only:
- RL : Size of object data in bits.

Common:
- H : Object's handle.
- BS : Size of extended object data, if any.
- X : Extended object data, if any.
- B : Flag indicating presence of graphic image. Only entities have this flag:
    ```
    if (graphicimageflag is 1) {
        RL : Size of graphic image in bytes
        X : The graphic image
    }
    ```
R13-R14 only:
- RL : Size of object data in bits.

Common:
- X : Object data (varies by type of object).
- X : Handles associated with this object.
- RS : CRC; The CRC includes the size bytes.

An example for the **Block** object:

Common Entity Data

Block name	T	2	
Handle refs	H		[Subentity ref handle (CODE 3)]
			[Reactors (CODE 4)]
			xdicobjhandle (CODE 3)
		8	LAYER (CODE 5)
		6	[LTYPE (CODE 5)]
			[PREVIOUS ENTITY (CODE 4)]
			[NEXT ENTITY (CODE 4)]
CRC	X	—	

And an example for the **Line** object:

R13-R14 only:
Start pt	3BD	10
End pt	3BD	11

2000 only:
Z's are zero bit	B		
Start Point x	RD	10	
Start Point y	DD	20	Use 10 value for default
End Point x	RD	11	
End Point y	DD	21	Use 11 value for default
Start Point z	RD	30	Present only if "Z's are zero bit" is 1
End Point z	DD	31	Present only if "Z's are zero bit" is 1, use 30 value for default.

Common:
Thickness	BT	39
Extrusion	BE	210
Handle refs	H	[Subentity ref handle (CODE 3)]
		[Reactors (CODE 4)]
		xdicobjhandle (CODE 3)
	8	LAYER (CODE 5)
	6	[LTYPE (CODE 5)]
		[PREVIOUS ENTITY (CODE 4)]
		[NEXT ENTITY (CODE 4)]
CRC X	—	

Object Map

The *object map* is a table that gives the location of each object in the file. This table is broken into sections. Each section consists of a list of handle/file loc pairs, and looks similar to this:

Set the "last handle" to all 0 and the "last loc" to 0L;
Repeat until section size==2 (the last empty, except the CRC, section):
    Short: size of this section. (Note this is in BIGENDIAN order, MSB first.)
    Repeat until out of data for this section:
        Offset of this handle from last handle as modular char.
        Offset of location in file from last loc as modular char. (Note that location offsets can be negative, if the terminating byte has bit 4 set.)
    End repeat.
    CRC
    End of section
End top repeat

Note that each section is cut off at a maximum length of 2032.

Unknown Section

This section is largely unknown; the total size of this section is 53.

Second Header

Consists of a beginning sentinel, followed by data, and ends with the ending sentinel.

Image Data (R13c3 and Later)

Consists of:

Start sentinel.
BMP and/or WMF data.
End sentinel.
Extended Entity Data
    EED directly follows the entity handle. Each application's data is structured as follows:
  |Length|Application handle|Data items|

Proxy Entity Graphics

Proxy entities (called "zombies" prior to Release 14) can have associated graphics data. The presence or absence of this data is indicated by the single bit called the "graphic present flag," which occurs mostly on entity-type proxies, and very few other entities.

Entity-type proxies are proxies where the related class itemclassid field is equal to 0x1F2. Valid entities are:

EXTENTS	Minimum and and maximum extents of the graphic
CIRCLE	Center and radius of circle
CIRCLE3PT	Three-point circle
CIRCULARARC	
CIRCULARARC3PT	Three-point circular arc
POLYLINE	
POLYGON	
MESH	
SHELL	
TEXT	
TEXT2	
XLINE	
RAY	

The following "Subent" items indicate changes for subsequently drawn items:

SUBENT_COLOR	
SUBENT_LAYER	
SUBENT_LINETYPE	
SUBENT_MARKER	
SUBENT_FILLON	Fill on if 1, off if 0
SUBENT_TRUECOLOR	Red, green, blue
SUBENT_LNWEIGHT	
SUBENT_LTSCALE	Linetype scale
SUBENT_THICKNESS	Thickness in z-direction
SUBENT_PLSTNAME	Plot style
PUSH_CLIP	3D clipping
POP_CLIP	Empty
PUSH_MODELXFORM	Transformation matrix
PUSH_MODELXFORM2	Transformation matrix
POP_MODELXFORM	Empty

For example, the format of a proxy graphic object is:

```
TEXT
        3 RD    start point
        3 RD    normal
        3 RD    text direction
        RD      height
        RD      widthfactor
        RD      oblique angle
        PS      text string, zero terminated and padded to 4 byte boundary
```

EOF

The DWG file terminates with the EOF flag, which indicates the end of the file.

THE FUTURE OF FILE FORMATS

In 2000, Martyn Day began a months-long investigation of the complex world of proprietary CAD file formats. His research was first published in a series of articles for the British CADdesk AEC *magazine. In this excerpt, he discusses the reasons why, with the next generation of solutions, data interoperability will worsen. The article includes an interview with Scott Borduin, the chief technical officer of Autodesk.*

The issue of CAD file formats goes beyond the basic "How do I get at my data?" question. It's at the very core of how the CAD industry works, and will work in the future. With CAD developers talking about "doing away with files" and developing Internet ASP-based solutions, your CAD models will become eminently more distributable. But how will this map to the longevity of your projects? The control of the application suite will be taken from your control and access to your company-sensitive information will be based on your ability to keep paying the ASP fees.

The maturity of the CAD market has been leading the industry to use proprietary file formats and customer lock-in as defensive weapons in their armories. The next generation of CAD systems, currently under development, will cause further data translation problems due to increasing complexity and proprietary technologies.

When I talk to the CTOs (chief technical officers) of the design tool developers, they seem to roughly agree in the areas that they are putting their R&D (research and development) budgets: XML (extended markup language), model servers, and providing project services — either via the Web or not.

Objects in CAD

Objects are a confusing issue, because it has two distinct meanings, depending on which side of the fence you happen to sit on:

- If you are an technologically aware architect, you probably relate objects to how the latest generation of CAD software allows you to create virtual buildings, modeling with walls, doors, and windows, instead of symbolic lines, circles, and arcs.
- If you work as a programmer at a CAD vendor, your first thought would be to the architecture and language in which the applications are written.

Both views are correct and, in terms of data exchange, have an impact on the way the industry is failing to communicate.

In an interview with Scott Borduin, Autodesk's CTO, I asked how developing CAD systems have been changed by the wholesale adoption of object-oriented (OO) programming methodologies, and what this has meant for the proprietary file formats.

Mr Borduin explains the impact:

> **The whole issue of information exchange in general is one of the most interesting things moving forward in our business.**
>
> **Part of the problem has to do with just the notion of the basic terminology "file format." At one time in the development of software, people would design the layout of what information was going to be inside a file, and then build code on top of that. I remember doing this myself! That notion of the file format being designed and central to the main code hasn't really been the case for quite a number of years, mostly because of the OO software revolution.**
>
> **It's to the point now that there is honestly no one in Autodesk who could tell you, byte by byte, what an AutoCAD, Inventor, or any of the other of our files actually looks like on disk. That has to do, first of all, with the nature of storing information in the OO software world. All that ever matters is that each individual object knows how to get itself back out of that stored environment. The system as a whole doesn't know what the object ordering is or what the development of each individual object is either. From the standpoint of being able to develop our software rapidly to meet customer's needs, that's a good thing.**

I'll never forget when I was working for another firm on a package called DDM. One of its restrictions was that it could only do eight colors. Every time it was presented to a potential customer, we would get hammered over the fact that our competitors could do 256 colors! So, we went back to the R&D folks and said, "You just have to fix this color thing and give us 256 colors!" They came back with an estimate that it would take approximately two and a half man-years of engineering to do this: they would have to find every single data read routine in the code and make it aware that the color field was going to be one byte wider!

So now it's a good thing that our storage format is now opaque to the programmer. We don't have to concern ourselves with system-wide details to make changes, but it does also mean that our storage format is opaque.

From Files to Central Databases

The introduction of OO programming to CAD developers meant that small teams could work on an engine like Autodesk's AutoCAD or Bentley Systems' MicroStation and develop tools without having to know the big picture. The benefit for users and developers alike is that development and new features now take a lot less time; systems develop quicker and offer richer feature sets.

CAD developers see files, however, as being a limiting factor. This one big lump of data isn't granular enough for developers, and requires complex management in a distributed enterprise.

So guess what? The next generation of design systems will aim to get rid of files and work from a central database that stores the data, serves up filtered information — all from a common model. There is, however, the downside.

Mr. Borduin describes how data exchange will become more complex:

> It's about to get worse, because the file is going to go away as the standard boundary of a database. We are going to have databases with more project-oriented structures layered on top of other back-end systems, like relational databases. From our perspective we won't know what the files are, what their extensions are, or what their locations are. And we won't care, because it will be someone else's software that's actually worrying about it.
>
> What's important now, in terms of getting information between systems, first of all is: what object model is each system working with? Secondly, what do people want to do with that object model?
>
> When we look at Autodesk's big SAP sales information system (which holds all our financial information) from the standpoint of the end user, the file formats and the location of servers are entirely irrelevant. The user doesn't care what the total scope of the information is. Instead, it's what do I want to know? Or, what do I want to edit or input right now?
>
> That's the way the CAD software is moving. We are going to end up with large architectures in which you are working on a specific subset of information from a project. Files are artifacts that you won't even see anymore. That raises the question of what does it now mean to do data exchange in the customer environment?

Working with Objects

Before looking at the implications of these large architecture, server-based systems, let's have a brief summary on modeling with objects in CAD — the walls, doors, and windows.

The idea is to move away from using the CAD software as an electronic drafting board, and move toward creating an intelligent, real-world model. You are no longer limited to lines, circles, and arcs, but you "draw" with doors, walls, and windows.

Technically, the doors, windows, and walls are all object classes, from which a wide variety of each type can be defined as each contains a number of variables and processes. For example, when you invoke the "wall" command in an object modeler, you are calling up the class definition wall. As you go through the process of entering in parameters — single or double-glaze, height, width, length — you create an object in the CAD database. When another object, like a window, is introduced to the model, the built-in behavior (process) allows the wall and the window (two objects) to interact in an architecturally meaningful way. Each real-world object is controlled by a small subprogram, called a parent application, which holds the behavioral intelligence to each instance of that object (single-pane window, double-pane window, etc.) in the model.

This is a major advance: from the use of drawings to merely represent, to the ability to model and analyze with intelligent feedback. The graphical interface still relies on vectors as its primary representation, so no big change there. The benefits can be seen, however, when performing insertion and editing commands. These objects will, if programmed to do so, interoperate with each other providing greater productivity. Inserting a window into a wall is point and click, with all the finishing details being handled by the respective parent applications, to your specification. If you want to move the window, simply grab it and move it; the wall repairs itself automatically.

The Downside to Objects

It all sounds grand in theory and to some extent this can be seen today in products like Autodesk's Architectural Desktop, Graphisoft's ArchiCAD, and Bricsnet's Architectural. The problem is, there are downsides. Walls, doors, and windows can do only what they have been programmed to do. They can be represented (i.e., displayed) only by how they are defined (or preprogrammed). The task of defining everything that goes into a building as an object, with the corresponding intelligence, is near Herculean.

Sometimes objects interact in ways you don't intend. When it comes to adding detail or custom features (be it in style or non-standard behavior), the systems can fall down. The file sizes tend to be on the large side.

When it comes to exchanging object information between different CAD systems, the problems compound. Object models include geometry, behavioral intelligence, annotated product information and relationships — not just the x,y,z-vector information, line styles, layers, and color that we are very familiar with, yet the industry couldn't solve the translation of either.

Today's Architectural Desktop software (ADT), for example, stores only object data in a DWG file, not the intelligence. Due to the way Autodesk has implemented its current ObjectARX language, non-ADT users have to download a separate file to enable AutoCAD and other vertical applications to edit ADT-generated objects. Competitive vendors have little chance of gaining access to the intelligence, other than via the still-incomplete IFC (industry foundations classes) format under slow-development by the International Alliance for Interoperability (IAI).

As this book is being written, CAD vendors are developing object modelers that come in the form of project-based systems. These offer to collate, store, and manage additional project-related data, which I'm sure will be embedded, hyperlinked, and cross-referenced with components (walls, doors, and windows) in the virtual model.

Somehow, in some way, the industry will have to be able to pass data from one company to another. Every CTO I have talked to admits that data exchange is going to get worse as these systems roll out. Bodies, such as the IAI and STEP, are developing cross-platform, independent, object-based solutions, but they typically take years, if not decades to complete.

Server-based Systems

CTOs from Autodesk and Bentley Systems concur that files are a parochial concept. At some point, files will be left behind. The word schema appears to be the substitute. It refers to a filtering or snapshot of a CAD model. It selects a predefined set of data from the model to suit the person who needs a subset of the model.

In this vision of CAD, a centralized model of the building (or mechanical part or map) is held in a database (typically provided by Oracle). The software serves up from the live data views of the model depending on the recipient's needs:

- An accountant gets costing per floor.
- The heating and ventilation contractor receives plans of HVAC components.
- The facilities manager interrogates inventory.

. . . all from the same file, all with access to different schemas.

While this is all technologically exciting, the CAD vendors are doing this for a reason. They see the data, which we in the engineering departments create, as having higher value throughout the remainder of our company. Unless CAD vendors go this route, it is hard to get that information distributed otherwise in a meaningful way. To a point, I agree.

By moving to a centralized project-base modeling environment, CAD developers see their technology becoming as important to their customers' businesses as SAP is today. The engineering data is turned into something with which nonengineering professions can do something. The hope is that, in moving from the engineering department's desktop to distributed information provider, CAD vendors get themselves a bigger slice of the total IT (information technology) budget.

I am sure there are large corporations looking for an all-encompassing solution like this. I can see, however, three immediate problems:

1. The **cost** of installing such a system, which requires big (read: expensive) servers, fat data pipes, and lots of consulting and customizing.
2. The **complexity** of centralized systems, which are making a comeback but are something from the 1970s that we ran away from with our independent desktop PCs.
3. The **dependency** on the fortunes of a single CAD vendor, who requires that we store all our project information in the vendor's proprietary model.

To access information, we rely on schemas to extract portions from the complete model. A single model may end up as seven or eight extracted files — HVAC (heating, ventilating, air conditioning), solid geometry, FM (facilities management), accounting, object instances, XML, etc. — that somehow need to be stuck back together at the other end. Frankenstein comes to mind. To achieve the perceived elegance and simplicity of a central server-based system, each corporation has to adopt an inherently closed and proprietary approach to storing critical engineering information. Each company becomes a digital island, albeit internally with improved communication of engineering data!

ASP-based Systems

In this Internet Age, the centralized infrastructure need not be physically in the company, but somewhere "out there." It's fashionable to outsource services, and model servers are one possibility. The engineering applications that generate the content (AutoCAD, MicroStation, et al.) reside in the engineering department, or can be served up live over the Web (with appropriately wide-enough bandwidth). Your data sits on a server somewhere on the planet (England, Hong Kong, or wherever the per-GB is cheapest), managed and serviced by an ASP (application service provider), who could be the CAD vendor itself — for an extra fee, of course!

I have, however, reservations about the ASP approach, because it could further complicate the data issue. You'll never own a physical copy of the software, only the right to use it. Your right to access your data is limited by your ability to continue making monthly or annual payments.

ASPs go out of business, and your outsourced engineering solution goes bust. ASP-based CAD software is automatically updated; changes to product may negatively alter your previous designs. (If you are designing a nuclear power station, you don't want the software to be developing out of your control).

I asked Mr Borduin about the issue of "going backwards to go forwards," and the problems of creating humongous corporate engineering systems. He predicts the future:

> **In some ways, it does look like "the good old days." On the issue of ASPs, that's a valid point. In the end, the customers own their data, so in any kind of ASP scenario the customer must have a reasonable assurance about their ability to get at that information — regardless of whatever end solutions they are working with, and by a variety of tools.**
>
> **Despite what centralization looks like, I think the problem is getting better. We are going to get to an era when the CAD application (this great big mammoth chunk of code) will interconnect with all these new architectures. It will force you into a multitier paradigm with swappable components between tiers. From the application standpoint, you will have a lot smaller subsets of things going on. You won't fire up just one CAD environment to add all the information, from geometry to detailing to cut lists and structural information. Every one of those will wind up being different small applications living at the top tier of all these architectures.**

It occurs to me that there is something fundamentally awry between the design software developers and the needs of their customers. While the customer's engineering projects may have a lifetime of 50 years or more, software developers exist in an environment that looks forward to the next three years. Here you have a dichotomy:

- **Clients** require longevity of dumb and intelligent data to mirror their projects, due to liability and facilities management requirements.
- **Software developers** feel they can't plan beyond a shorter timescale due to competitive pressures, the perceived demand for new features, and the very nature of the software industry, its tools, and constant development.

The Future of CAD Is . . .

It's fun to ask CAD developers where technology is going to lead. In my experience, the response consists of amorphous nonspecific answers, with words like "Internet," "distributed," and "concurrent" used liberally.

Dig deeper, however, and ask for specifics: What's in the next two releases? Or coming in the next 24 months? CAD vendors pretty much draw a blank.

There are two issues playing here:

- Vendors do not want to spoil quarterly sales figures by talking about the next software release before absolutely necessary.
- Perhaps more unsettling, they may not know what will be in the next software release.

Is your data safe in the hands of short-term thinkers?

Software companies do, however, care about the longevity and integrity of their customer's file formats. Extensive work is usually carried out with each new release to ensure compatibility with a couple earlier releases (called "backwards compatible). Occasionally, file formats are forward compatible (older software can read files created by newer releases). As software products mature, backward compatibility becomes more and more difficult because years of data baggage usually lead to a major schism at some point. This can prove to be an unnerving experience for customers with long-term projects and to CAD vendors who are expected to provide continuity. With the schism, the data must be mapped between systems and that, as we know, has inherent problems.

A case in point was Autodesk's development of Inventor, a mid-priced mechanical CAD modeling tool. It was developed to eventually replace its AutoCAD-based (DWG) Mechanical Desktop. Surprisingly, the first several releases of Inventor did not support the all-important DWG format. Mr. Borduin described the problem he encountered:

> **When we look at moving data between CAD systems, from my personal experience in trying to write the DWG translator for Inventor, it's very difficult, actually impossible to achieve 100 percent fidelity. Even when we had complete access to all proprietary information and source code, it was difficult.**

The Aging of Software

One solution is for the engineering department to freeze a complete system — hardware, software, and operating system — for the duration of a project. This, however, flies in the face of software upgrade policies, such as Autodesk's, who places a three-revision limit on upgrades (roughly five years). If you fail to upgrade an AutoCAD that's three revisions old, the software becomes unsupported, and you have to pay full price for an "upgrade."

The policy punishes customer by forcing them to buy upgrades they do not necessarily need; or seek special dispensation from the vendor.

Conclusion

It's proving difficult for CAD developers to win over the customer base of their competitors, to the point that unless something unfortunate happens (think Act of God, or United States Federal Commission deciding you are a monopoly). It's hard for me to see how the current market shares in mechanical CAD and AEC (architecture, engineering, and construction) are going to change. The battle for the right to sit on your desktop continues unabated by this however, although with more concentration into selling existing customers more technology and adjacent seats.

At this point in time there is a lot of opportunity for the technological betterment of design tools. I hope that CAD developers remember that their technology deployment decisions have long-term ramifications for their users' data and businesses. It's good to hear that CTOs are clear on the benefits of the next generation systems, and are aware of the problems and dangers they may create. It is, however, ironic that to move forward, they acknowledge we also have to takes steps backward. Perhaps there is no gain without pain.

To wrap up, I asked Mr Borduin what he would advise customers when looking at the future deployment of distributed CAD systems, especially if they are worried about legacy data and getting trapped.

He replied with the this advice:

> **If I were a customer, I would look very hard at the kind of information my company is putting in the design system today, and what information I think will be put in tomorrow. Then I would analyze where the data goes inside the organization, and try to identify the varying sets of information.**
>
> **This could be, for example, the plots sent to the bidders. To create these I need to be get cut lists. It's all about getting to finer granularity rather than thinking about the total database.**
>
> **Then I would look at each vendor's strategy to provide flexible open formats for getting out the information. I think that's the way the whole IT world is going today. Even though the technology on the CAD design side is always going to be extremely challenging and complex, nonetheless CAD is going to look like a specialized subset of IT.**

Hand drafted by Herbert Grabowski.

A Recent History of CAD

12

"There is a time for everything, and a season for every activity under heaven: a time to be born, and a time to die."
— *Ecclesiastes*

Companies, products, and technologies are conceived, born, mature, wither, and die. DOS, dBase II, and Lotus 1-2-3 are examples of products that dominated their market, seemed sure to be preeminent forever, and are now almost forgotten.

Technology seems to zoom forward into new, amazing, unpredictable areas. But there are recurring patterns:

- Late delivery of software
- The hype surrounding a promising new technology
- The renaming of lackluster programs
- Takeovers and bankruptcy
- Intense innovation, when there is competition
- Ebb and flow of personalities between companies . . .

CHAPTER SUMMARY

The final chapter of this book presents a recent history of CAD drawn from the pages of author Ralph Grabowski's *upFront.eZine*. This weekly e-newsletter has been published since May 1995.

Hence, this recent history of CAD starts in May 1995 and runs through April 2001, when this book was completed. All back issues of *upFront.eZine* are available at www.upfrontezine.com/welcome.htm.

RESOURCES

Joel Orr has put together a terse history of CAD since 1951 when EAI produced the first analog flatbed plotter at www.joelorr.com/caddhistory.htm.

A wonderful book recalling of the early days at Autodesk was written by co-founder John Walker. While his book, *The Autodesk File*, is no longer in print, you can download his manuscript in several formats from www.fourmilab.ch/autofile/.

Another brief, but more complete history of CAD from pre-1970 to 2000 is available at www.bozdoc.f2s.com/CAD-History.htm.

You read a few more details about Ivan Sutherland's Sketchpad at sun.com/960710/feature3/sketchpad.html.

The history of Autodesk and Bentley Systems is in Chapter 1 of *MicroStation for AutoCAD Users: A Bi-directional Guide"* by Frank Conforti and Ralph Grabowski (Delmar Publishers).

To understand why things are the way they are today, and to better comprehend the changes that will come in the future, it's important to know the past. The birth, adoption, and death of technologies and corporations described in this chapter help you evaluate hype from vendors.

Even though the history of CAD described in this chapter begins in May 1995, there were many, many events and trends prior to 1995 that determined the development of CAD. File formats developed 15 and 20 years ago are still going strong; file formats have an ability to transcend generations of hardware, software, and operating systems. Here is an example.

Bentley Systems got its start in 1986 writing software that could read, display, and eventually edit the files of another CAD program, Intergraph's IGDS (interactive graphics design system). Bentley first called their program PseudoStation, then renamed it MicroStation.

In the early 1990s, MicroStation became the first CAD program to read AutoCAD's DWG files without requiring the user to first translate the files. In the late 1990s, Pangaea created yet another CAD program, called DualCAD, that could read, write, and edit both MicroStation DGN and AutoCAD DWG files.

The Beginnings of CAD

CAD is now 40 years old. The concept of computer-aided design goes back to 1961, when Ivan Sutherland wrote his Ph. D. thesis at MIT describing the "Sketchpad." Two years later, he implements his theory in software running on a TX-2 computer, one of the rare interactive computers of the time. In the 40 years following, computer-aided design becomes a multibillion-dollar industry.

The events listed here are by no means complete, but rather give you a selective overview of what was happening in the realms of software, hardware, and operating systems over the last six years. You get a flavor of what life is like in the CAD community by the mention of corporate births, mergers, and deaths, along with descriptions of lawsuits and marketing efforts.

1995

The buzzword of the year is "object oriented" (OO, for short) and it seems every CAD vendor is adding the technology to their CAD system — or else is promising to. CompuServe continues as the dominant online service, and TAPcis software is the favorite way to access its email and discussion forums. For consumers, the Internet is largely unknown; I sign up in November 1994 with a local Internet service provider, the same month that Netscape is incorporated.

May 1995

Autodesk announces AutoCAD Release 13 for Windows 95 at the Comdex/Spring '95 show. When it ships, several months after Windows 95 itself, R13 features long file names, runs multiple sessions, supports OLE v2, and offers a multiplatform license for use on Windows v3.1, 95, NT, and MS-DOS. The CD-ROM includes separate versions for DOS, NT, and Windows v3.1. Around the same time, Autodesk ships AutoCAD LT Release 2, based on AutoCAD R12c4 for Windows.

The Vancouver AutoCAD User Society (the oldest AutoCAD user group) conducts a poll of 51 members attending a monthly meeting: 8 installed Windows 95; 3 installed Windows NT; and 8 installed AutoCAD R13, but 4 had gone back to R12. Autodesk admits sales of Release 13 are slow. CFO Eric Herr says the reasons are that: (1) the market hasn't moved to 32-bit hardware; and (2) customers are taking a long time to develop object-oriented applications. Release 13 goes into history as the least-liked version of AutoCAD. And with all the problems surrounding AutoCAD R13, Cimmetry Systems jumps the numbering system for its AutoVue viewing software from 12 to 14.

At one time, refrigerator magnets created concern because they could potentially erase data from diskettes. Oh, and AutoCAD hasn't run on Indigo computers for many years.

Bentley Systems starts showing off a new object-oriented file format planned for MicroStation "Version 6.0," which Bentley plans to ship in a year's time. Neither the OO file format nor "version 6," however, ever ship; the first major change to the DGN format does not occur for another six years. Bentley does ship PowerDraft, a lower-cost 2D-only version of MicroStation.

Intergraph unveils Jupiter, the code name for their object-oriented CAD system. When it ships, it is first known as Imagineer Technical; later, it undergoes a name change to SmartSketch.

Computervision, at this point still an independent CAD company, is shipping vertical applications based on their Polerus object-oriented toolkit. Other OO toolkits and CAD systems of the day include Cadkey's CODe (Cadkey Object Developer) and ART's Chief Architect.

June 1995

The computer common on desktops has a 40 MHz 386 CPU with 8MB RAM; more up-to-date users were fortunate to have a computer with a 60 MHz 486 and 16 MB RAM, while power users operate with a 90 MHz Pentium and 32 MB RAM.

The A/E/C Systems show has been the biggest event of the year for the CAD world since the mid-1980s. At this year's show, the phrase "object oriented" is on the lips of every CAD vendor, and on the page of every press release. The show seems slower, however, with an attendance of 20,435, down in a steady decline from its all-time peak of 28,788 (in Anaheim, California, 1989). In later years, AEC Systems desperately attempts to redefine itself, as attendance slips to 16,000. The Internet makes getting information at trade shows less relevant.

Basis Software releases an AutoLISP compiler called VitalLISP. There is some suspicion over the source of the software, because the company is located in Bentley Systems' hometown. It was known that Bentley had been working on an AutoLISP clone, whose development it had abandoned. Later, Autodesk purchases VitalLISP, and integrates it into AutoCAD 2000 as Visual LISP.

July 1995

Malcolm Davies becomes the head of German CAD vendor Nemetschek's North American operations. Mr. Davies was formerly the chief executive officer of Cadkey (where he cut the price of Cadkey software from $3,495 to $495, which saw sales increase but revenues remain flat), and prior to that vice president of Autodesk. Nemetschek is the #1 AEC CAD vendor in Europe.

August 1995

CAD vendors are cutting their prices. Intergraph reduces its $15,000 software to under $7,000; new products from Computervision are a much-cheaper $2,995; and Parametric Technology releases a lower-priced Pro/E Jr.

The second most common giveaway at trade shows is the pen — second-only to candy, which is costs even less.

Following the bug patches named c1, c2, c2a, and c3, Autodesk sends its customers even more bug fixes for Release 13. In total, users suffer through 351 bugs and nine bug patches in 12 months. Autodesk ships WorkCenter, which it later sells to Motiva, a company that goes out of business abruptly five years later, leaving customers stranded.

Vermont Microsystems wins a $25.5 million judgment against Autodesk over theft of trade secrets. (Autodesk had hired a VMI employee for his knowledge of display-list processing, which Autodesk had integrated into AutoCAD R13.) Upon appeal, VMI asked for $100 million, but the judgment is later reduced to $9.5 million. The court's decision makes fascinating reading at www.tourolaw.edu/2ndcircuit/july96/95-7279.html

Softdesk purchases Foresight Resources, whose DrafixCAD was the very first CAD package to run under Windows.

Newly independent of Intergraph, Bentley Systems is on a hiring frenzy.

September 1995

Intel announces the 686 CPU, which some pundits call the "Hexium" (because the 586 was called the Pentium). Intel eventually named the new CPU the Pentium II.

Xitron secures new financing, moves its offices to Arizona, and plans on shipping XCAD v3.1. At its launch a few years earlier, Xitron told a disbelieving media that it was aiming to sell one million copies of its low-cost 3D CAD software. The company never reaches its goal.

Autodesk purchases the South African CAD firm Automated Methods for its ReGIS mapping software. ReGIS is rereleased as AutoCAD Map. The company's other software, Ultimate CAD (an under-$1,000 2D CAD package), is terminated as part of the purchase agreement.

Spatial Technologies announces it is making its SAT (ACIS) file format publicly available. Over the following years, competition becomes so intense between Spatial and arch-competitor ParaSolid (owned by Unigraphics) that Spatial eventually sells ACIS to Dassault Systemes.

Softdesk ships Planix Home 3D, a software package easy enough for my 9-year-old son to use. He asks me for "that new game, the one where you can draw walls and pool tables."

The *CAD++* newsletter is the first CAD publication on the new Microsoft Network. The experiment is a failure for both *CAD++* and MSN; *CAD++* eventually becomes *upFront.eZine*, while MSN becomes a general purpose Internet portal. Bill Gates is quoted in

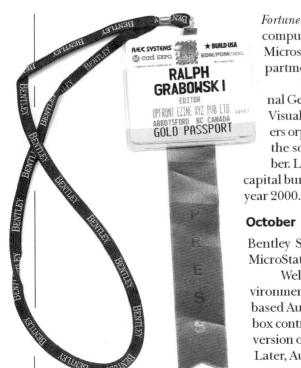

The show pass becomes a status symbol according to its color and the number of tags. Some attendees manage three tags: Press, Speaker, and Exhibitor. The holder sports the name of a sponsor, who pays the show organizers for the privilege of being around the neck of attendees.

Fortune magazine as still believing his fascist vision of "a computer on every desk and in every home, all running Microsoft software." Later, the United States Justice Department steps in to halt the vision.

Numera Software was formed by some of the original Generic CADD programmers. They hope to sell their Visual CADD software to the 300,000 Generic CADD users orphaned after Autodesk first bought, then shut down the software. Visual CADD v2 is due to ship in November. Later, Numera becomes an early example of "venture capital burnout" that becomes common with dot.coms in the year 2000.

October 1995

Bentley Systems ships version 5.5 of its CAD software as MicroStation 95 on November 16.

Wellcom Software of Germany has a programming environment for porting applications written with ADS (the C-based AutoCAD development system) and DCL (the dialog box control language) to MicroStation and Cadkey. The conversion of a complex application was said to take four weeks. Later, Autodesk phases out ADS with AutoCAD R14.

November 1995

Dan Raker's High Mountain Press sells its *MicroStation Manager* magazine to The MicroStation Community. Later, the magazine is sold to Bentley Systems. *CADD Times* magazine, which had catered to the under-$500 CAD market, closes down. Its editor, Randall Newton, later becomes the editor of *MicroStation Manager*.

3D/EYE begins shipping TriSpectives, the first ACIS-based 3D modeling software written specifically for Windows 95. One reviewer raves that this under-$500 software is a "Pro/E killer" but raves are not enough to save the product from eventual doom. Later, one portion of the company is sold to Autodesk, and the other portion to Visionary Design Systems. Autodesk uses the technology it acquired to create Actrix, while VDS creates a souped-up version of TriSpectives called IronCAD.

Supporters of VRML (virtual reality markup language) argue over the specs for version 1.1. At one time, VRML is seen as the 3D future of the Internet; even I get caught up in the excitement, and call VRML "the future of CAD." The prediction fails to come true, and the arguments slow down development of VRML to the point that 3D on the Internet never catches on.

Autodesk starts beta testing Mechanical Desktop, which is projected to be priced at half as much as Pro/E Jr.

December 1995

Corel begins beta testing CorelCAD for Windows 95/NT for release in February. The software is based on DesignCAD code licensed from American Small Business Computer. Later, CorelCAD is sold to IMSI and seems to disappear. As for ASBC, they announce a merger with Viagraphix, a training software vendor, who years later re-releases the CAD software as DesignCAD 3000.

Microsoft announces that its ActiveVRML will be released in summer 1996. The file format is envisioned as all-encompassing: a scripting language with sound, text, animation, and movies on top of 3D environments. When Microsoft proposes that ActiveVRML become VMRL v2, the VRML community revolts. Microsoft later drops its interest in VMRL.

Autodesk forms the secret Data Publishing Group, which creates a design and construction library of symbols. Later, the symbol libraries are sold to Thomas Publishing. Autodesk announces that it will start charging its third-party developers between $495 and $2,500 a year.

The PNG (portable network graphics) file format is proposed as a royalty-free alternative to the popular GIF format (developed by CompuServe). GIF was beset by royalty demands from Unisys, who owns the patent on the compression algorithm used by GIF.

At the end of the year 1995, I make predictions that:

- Autodesk will get AutoCAD R13 right. (True.)
- Operating systems will increase to 64 bits, with the Microsoft Cairo NT-based operating system due in 1998. (As of early 2001, this still had not come true.)
- The middleman will be eliminated. (Partially true.)
- Internet tools will begin to mature, such as FrontPage from Vermeer Technologies, later purchased by Microsoft. (True.)
- Microsoft's attempt to replace Java with Visual Basic will flop. (False.)
- Bentley Systems will grow stronger as an independent company. (True.)
- CAD drawings will appear on the Internet. (True.)
- Information will become a weapon for consumers when corporations and governments make misleading claims. (True.)

Although vendors still hand out buttons in a wide variety of shapes and sizes, few show attendees wear them — perhaps because we no longer wish to be associated with causes.

Trade show giveaways: keychain, letter opener, Post-It dispenser, and cardboard button.

1996

This is the year that CAD vendor first begin to embrace the Internet, albeit hesitantly. A survey I took at the end of 1995 of the Internet plans of 40 CAD vendors produced the collective reaction of "Huh?"

This year computing celebrated its 50th birthday. The ENIAC computer was first turned on Valentine's Day, 1946.

January 1996

Autodesk and Numera argue over who has the first Windows 95 logo-compliant CAD software. While AutoCAD was the first product to receive the logo, Visual CADD was the first to ship with the logo. Autodesk ships AutoCAD R13c4 on 22 January.

February 1996

The first sign of trouble for Visual CADD appears when Numera hands over manufacturing and distribution to Corel, and Corel reduces the price by $100 to $495. Novell closes down its Quattro Pro spreadsheet (acquired from Borland) after selling WordPerfect to Corel. Later, Corel purchases Quattro Pro from Novell.

Amid the deadline for proposals for VRML v2, Netscape says it is convinced "VRML is going to explode in a big way this year." VRML never becomes particularly popular.

March 1996

SoftSource is the first to ship a plug-in for Netscape Navigator that allows a Web browser to display DWG and DXF files. At the time, nobody cares; most CAD vendors question the need for such a product. Four years later, the displaying, editing, and collaboratively manipulating CAD drawings in Web browsers becomes the main focus of most CAD vendors.

Bentley Systems decides to concentrate on three markets: building/plant engineering, mechanical engineering, and geoengineering. It does so by bundling MicroStation with vertical add-on products.

Seagate unveils the first 23 GB hard drive ten years after 20 MB is considered plenty. By the year 2000, 20 GB hard drives become standard for home computers for storing MP3 music files and JPEG digital photos.

April 1996

Five years ago, CAD vendors had struggled to adapt their software to Windows. This year, they begin their struggle to do the same again for the Internet. This is the month that the A-B-C (Autodesk, Bentley, Cadkey) vendors tentatively stick their toes into murky Internet waters.

Autodesk announces DWF (drawing Web format), which goes on to become the basis of the Whip plug-in for Web browsers. Bentley licenses technology from Spyglass to add a Web browser to MicroStation. (Spyglass is the same company from which Microsoft licensed its Internet Explorer). Web browsers integrated into software, however, fail to become popular. Cadkey announces SiteSculptor, a shareware VRML authoring tool that uses 3D solids modeling and wireframe models; the product fails in the marketplace.

> **Also in April 1996**
> Apple fires ceo Michael Spinder, and replaces him with Gilbert Amelio.
> IMSI ships TurboCAD 3 for Windows 95.

May 1996

upFront.eZine comments: "We see this as a sign that the Internet is moving from being Major Media Hype to Just Another Tool for getting your work done."

More Web-related developments: Bentley releases a beta of its VRML import-export filter. Intergraph subsidiary InterCAP converts its 2D CGM (computer graphics metafile) to a Navigator plug-in. And industry observers are starting to question whether VRML has a future in CAD.

Numera muses over releasing Visual CADD as an enormous plug-in for Web browsers. This would allow users to view, edit, and plot Visual CADD drawings within a Web browser. The plug-in is released as a beta, but never ships formally. Nevertheless, the move foreshadows ASP (application service provider) software, where CAD programs run over the Web in browsers.

June 1996

In a stunning move, Cadkey, Inc. sells its Cadkey and CuttingEdge mechanical software to Baystate Technologies, but retains its DataCAD architectural software. (This event is the first of many mergers and buyouts between CAD vendors over the next several years.) Cadkey, Inc. changes its name back to its original name, Micro Control Systems. The company says it wants to focus on developing innovative 3D products for the Internet by converting its CODe CAD development system to JavaCODe. Nothing comes of the plan. Later, Baystate takes on the Cadkey name.

Nemetschek announces a Windows 95 version of their allPlan FT AEC software. Until now, the software ran on Unix and NT systems only, creating for market acceptance problems. The port isn't enough, however, to keep Nemetschek in the North American market.

> **Also in June 1996**
> TailorMade Software releases a DWF-to-DWG translator; Autodesk never releases such a translator.
> Corel starts shipping its 3D ACIS-based CorelCAD for $249.

> **Also in August 1996**
> Autodesk subsidiary Kinetix announces a 3D Studio plug-in for Netscape Navigator.
>   Pangaea Computer systems releases a DGN plug-in for Netscape Navigator, something Bentley refuses to do.
>   Eagle Point Software acquires SolidBuilder.
>   A Swedish group releases bCAD for Windows 95.

Bentley Systems continues to demonstrate Objective MicroStation 6.0, even though the product never ships. They show gee-whiz features like dragging a symbol from a Web site directly into the CAD drawing. Later, this becomes standard in most CAD software packages.

July, 1996

Parametric Technology Corp gets into the AEC market with the purchase of the object-oriented Reflex CAD of England. The product fails, and PTC changes direction toward the product collaboration market with its Windchill software.

Microsoft announces that Internet Explorer v4's user interface will become the future interface for all its Windows applications. While the browser interface indeed becomes the interface for future versions of Windows, it does not for most of its applications. Microsoft also announces that Windows 97 will ship in a year's time. With ensuing delays, the product is renamed Windows 98, and ships more than a year late.

Vincent Joseph Innocentius Everts, the flamboyant Dutch head of Cyco Software, announces that he will be marrying an American psychologist he met over the Internet. At 37, he thought he never would get married.

August 1996

The VRML v2 specification is released at the Siggraph show, but has taken so long to be developed that few care anymore. The VRML organization has since become the Web 3D Consortium at www.vrml.org.

Most mousepads are rectangular and covered with cloth. But not all: some are round, others are square, and some sport a pebbled plastic finish said to be better to tracking with a mouse ball.

IDG launches *JavaWorld* as an online magazine, which continues to this day at www.javaworld.com. While Java doesn't take over the world, it slowly and steadily makes a place for itself in the background of the Internet.

Baystate Technologies announces that Cadkey 97's price will increase from $795 to $1,195. Later, the price will further increase to $2,195. From that experiment in pricing, CAD vendors understand that low prices do not equate into higher revenues.

September, 1996

Users find out the hard way that Numera has burned through all its investment money. Numera's Web site goes offline, and its 800-number is disconnected.

The US Department of Justice begins to take a look at how Microsoft conducts its business. Meanwhile, in Burma, ownership of a modem or fax machine is made a crime punishable by 15 years in prison.

Autodesk announces "its most significant wave of Internet products to date" at its Autodesk University trade show being held in Chicago:
· Internet Publishing Kit for embedding URLs in drawings and saving drawings in DWF format
· Internet version of their PartSpec library
· Java-enabled Whip plug-in for Web browsers
· WorkCenter document management software for the Web
· And Autodesk purchases MapGuide for placing maps on Web sites

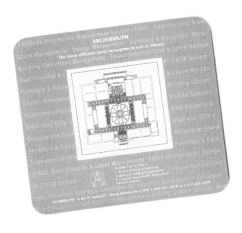

October 1996

The managers of the DataCAD Product Group form a new company, DataCAD LLC. They purchase the DataCAD software from Micro Control Systems (a.k.a. Cadkey, Inc.).

To help improve sales of AutoCAD Release 13, Autodesk dusts off the "176 reasons" campaign it had utilized to launch AutoCAD R12. The ad copy reads, "For all those who have been asking the question, 'Can AutoCAD Release really make my life easier?', we offer 176 yes's [sic]." Keen CAD users pour over the list, and soon discover problems with the 176 "new" features. Four reasons duplicate each other, while another 30 features were already available in Release 12. The net total is 142 new features.

November 1996

Corel acquires the source code for Visual CADD from Numera. Later, the Generic CADD-compatible code is sold to IMSI (maker of TurboCAD), then on to TriTools Partners.

Baystate launches a lawsuit against Bentley Systems and a third-party developer who reverse-engineered Cadkey's PRT file format. Specifically, Baystate is concerned about the wording used in certain data structures of the translator. Baystate eventually loses the suit.

Bentley announces Engineering Back Office (later renamed ModelServer Publisher), a $24,500 multichannel electronic publisher of engineering data, and ModelServer Continuum, a $37,500 server that creates a contiguous database of engineering and enterprise data.

Owen Wengerd releases CADlock, software that locks an AutoCAD drawing to make it read-only. The software becomes popular for allowing clients to view and plot AutoCAD drawings, but not edit them.

This "CAD Cafe" mug was handed out as a promotional item by the now-defunct Autodesk Retail Products group located near Seattle, Washington. The group was responsible for Generic CADD, as well as non-CAD software.

Autodesk announces that all future versions of AutoCAD will be written only for Windows 95/97 and NT. Later, Autodesk changes the AutoCAD numbering system to match Microsoft's. Meanwhile, Bill Gate's new vision is that the computer "desktop" is the Internet. Synet sues Microsoft over the "Internet Explorer" trade name.

Rumors fly that IBM . . . , no Hewlett-Packard . . . , no Microsoft plans to buy Autodesk for $1 billion. None of the rumors are true.

December 1996

Following several weeks of rumors, Autodesk and Softdesk announce a merger agreement valued at $72 million worth of shares. The price goes up after PTC makes a counteroffer.

Autodesk announces that AutoCAD Release 14 will ship in July, 1997. And it raises the dealer price of AutoCAD by $300.

The editor of *MicroStation Manager* magazine calls on Bentley Systems to "embark on a campaign to convert some of those dissatisfied [AutoCAD R13] users." Instead, Bentley targets Cadkey users after winning the lawsuit brought against it by Baystate. With the lawsuit out of the way, Bentley announces its "Cadkey Migration Package" that bundles PowerDraft, Draft-Pak Mechanical, Cadkey Importer, QuickVision, Engineering Links, and one year of support for $595.

Also in December 1996

Russian software company Top Systems releases T-FLEX CAD, a 3D modeling system.

ZD Labs unveils a new benchmark that uses MicroStation 95 as its CAD component, partly because that version of MicroStation was available on thirteen operating systems.

Samsung Electronics says it will ship 1 Gbit DRAM chips in 2005.

Some mugs have great art; other mugs serve strictly as a vehicle for a corporate marketing message.

1997

upFront.eZine makes the following predictions for 1997:

- ActiveX Desktop will fail. (True.)
- Autodesk will move to a subscription model. (True but not a roaring success.)
- The PalmPilot will become a big deal. (True.)

And the newsletter suggests that readers keep an eye on the following software packages:

- AriTek's beta of Builder's Sketchpad, software that designs a house by snapping together rooms. The software doesn't get much beyond its beta status, but foreshadows numerous other easier-to-use architectural CAD packages, including one that becomes the basis of The Sims game. Later, AriTek is spun off from Softdesk, becoming an independent company for the second time.
- Straight Forward Software releases Plan Master LT, an architectural add-on for AutoCAD LT. The product is significant because Autodesk had deliberately hobbled LT's ability to run vertical applications. This did not stop clever programmers who eventually develop an entire programming environment for LT, as well as numerous add-ons.
- After five years of development, SoftSource releases Vdraft (short for Virtual Drafter). It had been code-named "Project Sausalito" after the city where Autodesk's headquarters were located at the time. This product was significant for being the first non-Autodesk CAD program to use DWG as its native file format. The software enjoys modest success, and spawns a new niche industry of DWG-based CAD packages. Others include the German FelixCAD, the American IntelliCAD, and the Canadian DualCAD.

January 1997

Graphisoft sues Graphsoft over the similarity of its name. Graphsoft agrees to change its name to Diehl Graphsoft. Later, the company is purchased by Nemetschek and renamed Nemetschek North America.

FIT (Furukawa Information Technology) ships Cadvance v7, and increases its price to $1,990 — another CAD vendor who finds it cannot survive selling software at $495. Cadvance had been the first major CAD package to switch from DOS to Windows. At the time, I had called it "the future of CAD." While

---

Also in January 1997

Bentley Systems announces MicroStation Modeler, its first mechanical CAD software.

Visio Corp. ships Visio Technical v4.5 for $299, the first CAD-oriented dialect of Visio. Later, Microsoft drops the Technical version with Visio 2002.

The AEC Systems trade show is sold to Penton Publishing.

Miller Freeman, publisher of *Cadence* magazine, had been producing the Autodesk University trade show for Autodesk, but now declines to rebid for the next contract.

Windows did indeed become the future of CAD, Cadvance itself did not, and lags in adding the latest Windows-oriented features.

Autodesk changes the name of its ARx programming interface (AutoCAD runtime extension) to OjbectARX to emphasis its object-oriented nature. Autodesk ships a consumer software product called Picture This Home! Kitchen; later, Autodesk later sells the software.

February, 1997

Convergent Group announces it will no longer invest in the software development of its GDS v5.6 and MicroGDS v5.1 CAD software. Later, MicroGDS is taken over by Informatix of England.

3D/EYE is sold in parts. One part is purchased by Visionary Design Systems (VDS), and the other by Autodesk. VDS uses the TriSpectives technology to create IronCAD, and later changes its name to Alventive. For its part, Autodesk uses the technology to create Actrix, an unsuccessful competitor to Visio.

Baystate announces Cadkey 97 will ship with integrated solids modeling. The company appeals its court loss to Bentley Systems, saying that it believes data structures must be protected under US copyright law. The company loses the appeal.

Also in February 1997
IMSI announces it has licensed ACIS for future CAD products.

Corel releases WordPefect Construction Edition for the construction industry, which bundles Visual CADD, WordPerfect, Quattro Pro, and 400 house plans.

The state of Texas begins to investigate Microsoft's anticompetitive behavior.

R13 versions of DWG translators finally begin to appear.

Kinetix releases the Hyperwire 3D authoring tool for the Web; the product was originally called Opus.

Intergraph launches Imagineer v2.

A sheetmetal-encased notepad handed out at a press conference by e-IDC.com.

March 1997

Visio Corp pays $6.7 million for the source code and nine employees of Boomerang Technologies. The source code consists of an AutoCAD clone that eventually becomes IntelliCAD. The history of IntelliCAD is complicated, but also one of the most fascinating in the history of CAD.

Several years earlier, a company by the name of IntelliCAD developed a software product called AutoCAD Data Extension. ADE made it easier to extract data from AutoCAD drawings. The software is eventually sold to Autodesk.

Softdesk purchases IntelliCAD, the company. Shortly thereafter, Softdesk asks its staff to undertake a secret project to cre-

> **Also in March 1997**
>
> The rumored purchase of Netscape by Oracle turns out to be false.
>
> Intel says it was able to run its Pentium CPU at a scorching speed of 451 MHz with an icepack.
>
> Rumors fly that Windows 97 will be renamed 98.
>
> And software vendors attempt to convert their products to Java, such as Java-based WordPerfect; in many cases, the efforts fail.

ate a clone of AutoCAD, after Softdesk's management is spooked by the idea that Autodesk might cut off access to the AutoCAD developer program — as Autodesk had done to Cyco Software. Only a few of Softdesk's executives and the ex-IntelliCAD employees know about the "skunkworks" project. Called Project Phoenix, the software is a clone of AutoCAD that reads and writes DWG and DXF files, has most of the AutoCAD command set, and includes compatible subsets of AutoLISP, ADS (AutoCAD development system), and DCL (dialog control language).

After Autodesk purchases Softdesk for its ObjectARX-based architectural and civil engineering software, Autodesk discovers it unknowingly also purchased the secret AutoCAD clone. A competitor to Autodesk tips off the United States Federal Trade Commission, who tells Autodesk to: (1) get rid of the IntelliCAD technology; (2) not attempt to reacquire Project Phoenix; (3) not attempt to acquire any company that owns or controls Project Phoenix; and (4) not prevent Autodesk employees from moving to the competing company. The controls are in place for ten years. The newly-independent Project Phoenix employees take the code and form a new company called Boomerang, located in San Diego, California.

The War of the T-Shirts, Part I: At the A/E/C Systems show where Visio first showed its "Phoenix Project" behind closed doors, members of the media were given black T-shirts reading "I've seen the future of CAD" (front) and "but I'm not allowed to talk about it" (back).

Shortly after the FTC injunction, Visio Corp. purchases Boomerang. Work continues on developing the AutoCAD clone, now named IntelliCAD. Visio's plan was to undercut Autodesk by offering IntelliCAD for $349 — ten times cheaper than AutoCAD and with 90 percent of the features.

The plan fails, and Visio Corp eventually gives the source code of IntelliCAD 2000 to an independent group called the IntelliCAD Technical Consortium. It is rumored that the IntelliCAD experiment had cost Visio Corp. $25 million.

Autodesk launches its VIP subscription program at $245/year for AutoCAD. Over the next several years, Autodesk repeatedly changes the terms of the subscription program, creating confusion among users. For a time, the program is limited to firms owning at least five Autodesk products.

Autodesk ships 16,000 beta copies of AutoCAD R14, which ends up becoming the most popular version of AutoCAD. When Autodesk does begin shipping R14, it does so nearly two months earlier than planned. Autodesk also begins shipping Mechanical Desktop v1.2 for AutoCAD R13 for $6,250.

Intergraph releases Solid Edge v3. Later, the product is sold to Unigraphics Solutions as part of Intergraph's attempt to regain profitability.

> **Also in April 1997**
> Rumors that America Online will buy CompuServe turn out to be true.
>
> Oracle decides not to buy Apple Computer.
>
> Daratech finds that MicroStation outsold AutoCAD in the Windows NT market during 1996; Bentley, however, fails to keep its lead.
>
> Microsoft says that the PC in 1998 should have a 200MHz CPU, 32MB RAM, and only PCI buses (no ISA buses).

April 1997

Intergraph sues Bentley Systems to see the latter's accounting books. Intergraph continues to announce net losses, but still manages to end up as #928 — and the only CAD vendor — on the Fortune 1000 due to its $1.1 billion in revenues.

ComputerVision announces its CADDS 5 parametric solid modeling software for Windows for $3,500. The company continues to lose money, and is eventually purchased by Parametric Technology.

May 1997

Graphisoft shows off ArchiCAD for Teamwork, which allows a team to individually work on portions of the same building project. At the time, there is little interest in the ground-breaking product. But by 2000, other CAD vendors scramble to make their software team aware, and even Microsoft gives away its NetMeeting software for free.

> **Also in May 1997**
> Intel gets into trouble for not taking quick enough action on the infamous floating-point error in Pentium CPUs.
>
> Borland sues Microsoft over employees Microsoft hired away for their expertise in writing programming tools.
>
> Kinetix ships 3D Studio VIZ for the architecture and automotive market at $1,995.

June 1997

Digital Equipment sues Intel for misappropriating patents used in Digital's Alpha CPU. Intel countersues, saying Digital failed to return confidential data. The next day, Cyrix accuses Intel of copying additional patents.

Visio Corp licenses MarComp's DWG/DXF read-write library for use in its Visio Technical software. Later, Visio buys the company.

Four CAD dealerships merge into a single super-dealership. CAD-PRO Systems Integration, CADworks, NECAD, and Premier Design Systems form Avatech Solutions. Later, Avatech buys up additional dealerships in the United States becoming the second-largest AutoCAD dealer in the world.

The vote for an Imagineer newsgroup fails after the votetaker declares "rampant fraud." Nearly half the votes cast were invalid because they represented repeated votes. Intergraph must wait six months before trying again to form the newsgroup.

Bentley Systems announces that MicroStation 95 will be the last "monolithic" release. Users will instead receive quarterly update CD-ROMs via its Select subscription service. The prediction proves untrue when Bentley releases additional monolithics: MicroStation SE (select ensemble), J (Java), and V8 (version 8). Bentley hopes to convert all of its users to its Select subscription service; while more than half eventually sign up, Bentley finds that a major release is still required from time to time.

Nemetschek admits that selling its allPlan software into the North American market has been a tough haul for the last two years. Instead of fighting Autodesk, the plan now is to latch onto AutoCAD. Nemetschek rewrites allPlan as a series of ObjectARx modules (called ARCH 14) that work with AutoCAD R14. The revised plan of attack ultimately fails.

A keychain puzzle handed out by Bentley Systems at tradeshows to celebrate the launch of its ww.Viecon.com Web site.

Also in July 1997

America Online becomes the largest ISP (Internet service provider) in the world.

Apple announces the resignation of CEO Gilbert Amelio; speculation is that Steve Jobs will return to lead the company.

John Lynch resigns from Autodesk as chief technical officer.

July 1997

More acquisitions: France-based Dassault Systems, the largest CAD company in the world, acquires SolidWorks for shares valued at $310 million. 3Com completes its acquisition of US Robotics, who had earlier acquired Palm Computing. Later, 3Com spins off Palm Computing as an independent company.

Bentley Systems purchases *MicroStation Manager* magazine from The MicroStation Community.

Computervision ships a boatload of software, including CADDS 5 Revision 7, Medusa NG, Optegra, and several packages prefixed EPD (electronic product development).

Industry observers wonder about Microsoft modifying Java to allow direct calls to Windows, thereby creating an incompatible version of Java. Later, in 2000, Sun Microsystems wins a law suit against Microsoft, but by then Microsoft has created a Java clone called C#.

August 1997

Baystate Technologies purchases FastSURF.

Intel offers $50 to anyone who can prove they were inconvenienced by Intel's misrepresentation of the speed of some CPUs.

September 1997

SDRC wins its appeal against Ashlar's patent infringement suit. Ashlar had alleged that SDRC's software infringed on Ashlar's user interface patents for its Vellum software. The win is important, since Ashlar had sued numerous other CAD vendors, most of whom paid Ashlar royalties.

Corel exits the CAD market by selling all its CAD software. IMSI, the maker of TurboCAD, buys CorelCAD 3D Modeler (formerly CorelCAD), CorelCAD Technical (formerly Visual CADD), Corel Personal Architect (never released), and other software for $5.6 million.

October 1997

In a complicated deal, EDS spins off Unigraphics Solutions as a new joint venture company it forms together with Intergraph. The new company, called UGS for short, takes over the SolidEdge software from Intergraph. From EDS, UGS gets the Unigraphics CAD software, the ParaSolid solid modeling kernel, and the iMAN product management software. The move allows UGS to aggressively compete ParaSolid against ACIS from Spatial Technology, resulting in an explosion of features and releases over the next three years from both vendors. Later, Unigraphics Solutions becomes completely independent, and changes its name to UGS.

UGS says it will remove the ACIS modeler from SolidEdge and replace it with ParaSolid. Bentley Systems announces it will do the same for its MicroStation Modeler. Other CAD vendors, however, implement both solid modeling kernels, ACIS and ParaSolid.

In its SEC filing, Eagle Point Software reports that it has decided to partner with Visio on the IntelliCAD product. The company can no longer sell AutoCAD as a retailer because Autodesk is concerned about IntelliCAD's $349 price. Eagle Point continues to be an AutoCAD third-party developer. Visio announces that IntelliCAD beta will ship in a month's time.

Also in September 97

The DataCAD AEC software is being used by 150,000 users.

FelixCAD directly reads and writes DWG files.

MicroStation SE no sooner enters beta than Bentley Systems announces a future version of MicroStation that can be customized with Java.

Windows 97 is officially renamed 98, and delayed until mid-1998. Microsoft declares it will be the end of the line: "After Windows 98, there will be only one Windows for the desktop, and this will be based on NT." Windows 98, however, is followed by 98 Second Edition and Millennium Edition (Windows ME). The merged operating system won't appear for another four years, as Windows XP. Microsoft also talks up DNA (short "distributed network architecture"), which fails and is later replaced by the .Net initiative, which, at the time of writing this book, is two years away.

The Federal Trade Commission begins to look into unfair trade practices by Intel, which controls 85 percent of the CPU market. Later, Intel agrees to conditions set down by the FTC. By then, AMD has become a strong competitor. In 2000, AMD is the first to ship a 1 GHz CPU for desktop computers.

November 1997

Computervision introduces the $3,750 DesignWave software as their next-generation assembly-centric architecture. The product is Computervision's first since becoming a software-only company, and the first mechanical CAD software to include Microsoft's VBA (Visual Basic for Applications). One week following the announcement, PTC buys Computervision for $260 million in shares. The Computervision name, one of the earliest CAD vendors, disappears.

Bentley begins shipping MicroStation SE. The marketing campaign employs a triple entendre: "14 Reasons to SELECT MicroStation SE." The "14" refers to AutoCAD Release 14; "SELECT" refers to Bentley's subscription program; and the entire phrase refers to a list of 14 reasons to select the updated software. Bentley Systems offers MicroStation SE to AutoCAD users for a price of $495.

At a conference in Rome, Bentley Systems announces that its forthcoming MicroStation/J will contain a native AutoCAD editor written in Java. The announcement fails to come true, and Bentley makes a similar promise for MicroStation v8. Users are more concerned, how-

---

Also in November 97

Intergraph sues Intel for coercing Intergraph into giving up patent rights to Intel. The law suit drags on for years.

The first CAD software appears for handheld PCs.

Autodesk purchases Vital LISP from Basis Software; later, the software renamed Visual LISP and added to AutoCAD 2000.

Thomas Publishing acquires libraries of Autodesk Data Publishing unit.

ever, that MicroStation/J will run on Windows operating systems only; the irony is that Java is meant to run on all operating systems.

Visio Corp. and SolidWorks enter into an agreement to exchange information on each other's file formats. Visio plans to give its Visio Technical and IntelliCAD products access to 3D solid models created by SolidWorks. For Visio, the plan never comes true, but in 2001 SolidWorks gains the ability to embed and edit Visio diagrams.

IMSI announces that TurboCAD 5 Professional will contain TurboLISP and ARxADS, which allow the CAD software to access AutoLISP, ADS, and ObjectARX program written for AutoCAD. The announcement becomes partially true when the development effort ends due to IMSI's poor financial health; a partial API is included in TurboCAD. ISMI announces that TurboCAD Solid Modeler v2 will be the new name for the recently acquired CorelCAD 3D.

AutoSketch makes a reappearance as the $99 Release 5, merging elements of AutoSketch with Drafix CAD that AutoCAD acquired as a result of its purchase of Softdesk. AutoSketch had originally been released in 1987 as a $80 CAD package to prove that Autodesk could write a low-end CAD program.

And the first Y2K (year 2000) scare stories begin to appear in the media.

December 1997

Controversy swirls over the winner(s) of the Architectural CADD Shoot-out sponsored by the Boston CAD Society. The marketing departments of the CAD vendors manipulate the complicated results to show they had won. Organizer Geoffry Moore Langdon definitively states that the MiniCAD team won the Architectural CADD Cup . . . but that allPlan was the favorite of the jury . . . but that the ArchiCAD team was the medal winner . . . but that MicroStation TriForma was the favorite of the audience. So valuable is winning that one CAD vendor threatens a lawsuit to be declared the winner.

Also in December 1997
TurboCAD v4.1 is able to read and write MicroStation files, in addition to AutoCAD and 3D Studio.

PTC ships Pro/E R19.

Autodesk ships Mechanical Desktop 2.

HTML v4 is approved by the World Wide Web Consortium.

Graphisoft says that ArchiCAD will integrate the IAI's Industry Foundation Classes.

Corel begins to show quarterly losses that won't stop until 2001.

With the cost of CD-Rs under 50 cents, they become a popular way to distribute demo versions of software, press kits, and images of award winners. CDs don't always have to be round to work.

1998

This is the year that Visio Corp. repeatedly makes the headlines as it acquires, then divests itself of AutoCAD-compatible technology. And this is the year companies start getting serious about the Y2K "bug" after a report says the software programming shortcut could cost the economy $600 billion.

January 1998

Visio Corp. purchases MarComp and its DWG-DXF read-edit-write API. Licensees of Marcomp's libraries worry that Visio will cut them off. Visio says it will honor all of Marcomp's existing licenses with competitors, which it does.

Microsoft's marketing takes aim at the Palm with no effect. In turn, 3Com sues Microsoft over the use of the PalmPC name. Later, the FTC investigates misleading "Can your Palm do this?" ads run by Microsoft and Hewlett-Packard. Products based on Microsoft's Windows CE (called "Wince" by critics) fail to get a significant market share.

February, 1998

In a stunning reversal, Visio Corp hands its recently acquired MarComp DWG APIs over to the OpenDWG Alliance. Visio creates the alliance, gives it its initial funding, and provides it with employees. The alliance's job is to make the APIs freely available for noncommercial and in-house use; commercial licenses are $5,000. Fifteen CAD vendors each pony up $25,000 to become founding members. The catch to membership is that you have to return to the alliance everything you learn about the DWG file format. (The format has never been publicly documented, even though Autodesk is fond of calling DWG "the standard" for CAD drawings).

Autodesk sees no benefit in joining the alliance. The company mocks Visio for creating the organization because, as Autodesk sees it, this is an admission that IntelliCAD is not 100 percent AutoCAD-compatible after all; Visio needs help from the rest of the CAD world to figure out fully the DWG format. Further, Autodesk accuses some of the founding members of hypocrisy because their file formats are closed, such as Visio's own VSD file format.

MarComp licensee Bentley Systems initially refuses to join, but later changes its mind. Three years later, the alliance is unable to fully document the R13, R14, or R2000 DWG formats.

Also in January 1998

Bentley Systems acquires Jacobus Technology.

Intergraph Computer Systems, manufacturer of hardware systems, begins operating as a company separate from Intergraph.

Autodesk sells its WorkCenter software to Motiva, and makes an investment in the company.

Softdesk (originally called DCA Engineering) founder Dave Arnold resigns from Autodesk.

Stefan Dorsch writes an ActiveX Automation interface for AutoCAD LT.

The official International TurboCAD Users Group launches.

Cadmax ships version 8 of CADMAX.

March 1998

The OpenDWG Alliance blows $371,000 (most of its initial funding) on a single full-page ad in the *Wall Street Journal* attacking Autodesk: "When it comes to your company's CAD drawings, Autodesk has you right where they want you," reads the headline.

April 1998

On a Monday, Visio Corp. ships IntelliCAD 98, late. On the following Friday, Visio fires the IntelliCAD chief architect for failing to include features Visio had wanted, such as associative hatching. The following Monday, ten members of the IntelliCAD team quit in protest. Visio recovers by quickly hiring a new team of programmers.

Autodesk is rumored to be working on a "Visio killer" in response to Visio shipping IntelliCAD, its $349 "AutoCAD killer." At the AEC Systems show in June, Autodesk shows off Actrix, its answer to Visio. Ironically, both IntelliCAD and Actrix do poorly.

> **Also in March 1998**
> SolidWorks ships SolidWorks 98.
>
> Dassault Systemes previews CATIA 5.
>
> Unigraphics ships Solid Edge v5.
>
> Nemetschek renames ARCH 14 as Palladia X along with the slogan, "The New Paradigm: Architecture the Way It Should Be."
>
> Apple kills off its Newton personal digital assistant, which has become the butt of many jokes for its poor ability to read handwriting.
>
> Microsoft announces that Windows NT 5.0 will be delayed until mid-1999; later, it is renamed Windows 2000, and fails to ship until early 2000.

The War of the T-Shirts, Part II:
At the A/E/C Systems show, where Autodesk employees (wearing sunshades) first show its new Actrix software, members of the media were given black T-shirts reading "The secret's out" (front) and "Actrix Technical" (back).

Visionary Design Systems uses Stalinist art to market its IronCAD software.

The other purchaser of 3D/EYE technology, Visionary Design Systems launches IronCAD for $3,995. The new midrange solid modeler is based on 3D/EYE software that VDS had earlier purchased. VDS uses Stalinist-style graphics to market the software.

The Federal Trade Commission launches a probe into Autodesk, but finds no problems. *Computer Reseller News* speculates that the OpenDWG Alliance sparked the investigation. While the investigation is underway, Autodesk's CEO declares that Autodesk has a mere 4 percent share of the CAD/CAM market, a number much smaller than the 70 percent usually bandied about.

May 1998

This month, Douglas Engelbart is honored for creating the mouse 30 years ago while at Xerox's Palo Alto Research Center (PARC) in California. He was also the inventor of multiple on-screen windows, a technology used today by almost all computers.

Autodesk throws the OpenDWG Alliance a curveball. DWG R14.00 included a "watermark" feature, which was not turned on until R14.01. The watermark reports whether AutoCAD is the software that created the DWG file. The change causes IntelliCAD 98 to crash, until Visio is able to ship a fix a few days later. Autodesk, perhaps mindful of the ongoing FTC investigation, comes close to making an apology by posting a patch to R14.01 that cures the problem in MarComp-based CAD software.

After two years of development, Autodesk announces it will ship Architectural Desktop (ADT), its single building model add-on to AutoCAD, for $1,395 during the summer. Two weeks prior to the announcement, Bentley Systems sends an email to industry editors giving

---

Also in April 1998

Autodesk ships AutoCAD Mechanical, a 2D-only mechanical design add-on for AutoCAD.

Kao, at one time the world's largest manufacturer of floppy diskettes, announces it will no longer make them.

eight reasons why its TriForma product is superior to ADT. Autodesk's reaction is, "If you know how Bentley got hold of NDA (non-disclosure agreement) software, I'd like to know." Later, ADT receives a volley of attacks from other CAD vendors.

Autodesk demos "Rebicon" at a dealer conference in France. Later, the product ships as the highly successful Inventor for 3D mechanical design, and industry observers wonder if the Inventor code will become the basis for a replacement to ADT. The speculation is unfounded.

At its annual Visio conference, the company announces it has sold two million copies of Visio software, but does not disclose the number of IntelliCADs sold. Later, when the number is revealed, it proves disappointingly low. Visio Corp. announces its goal to become the single worldwide standard for business drawing. Visio managers talk about a path of convergence for Visio and IntelliCAD, but are vague on the details; the plan is later abandoned, although some IntelliCAD features find their way into Visio. Third-party developers announce add-ons to IntelliCAD. Visio "6.0" is previewed and is said to ship in 1999; later, renamed Visio 2000, it ships in 2000.

CAD vendors go manic with acquisitions. Parametric Technologies acquires ICEM Technology, a division of Control Data. SoftTech acquires Adra Systems, a division of MatrixOne. Bentley Systems acquires Macao technology from Group Scetauroute. And Autodesk acquires Genius-CAD Software for $68 million.

Roopinder Tara leaves his position as editor-in-chief of *Cadence* magazine, and turns up at IMSI. Later, he leaves IMSI to form the highly successful Tenlinks.Com Web portal for CAD users.

Intel announces that Celeron will be the name for its new home-market CPUs, while Xeon is the name for new server and workstation CPUs. The Celeron becomes very popular due to its low cost, but the Xeon languishes when its performance isn't high enough to justify its premium price.

Also in May 1998
Ashlar ships a 2D-only version of Vellum called Draft 98.

DataCAD LLC ships DataCAD 8, the first 32-bit version of that software.

IMSI prices its ACIS-based 3D Modeler v2 at just $99.

Wang Global loses a court case in which it claimed it had a patent on the Web browser.

Borland changes its name to Inprise; later, it changes its name back to Borland.

Apple announces that its Unix-based OS X will ship in the fall of 1999; the operating system fails to ship until Spring 2001.

Wired magazine is sold for $65 million.

June 1998

At the AEC Systems '98 show, the first Web-enabled applications for CAD are shown. The Internet, however, still has not reached the consciousness of CAD marketing departments. Instead, high- and low-end CAD vendors alike are spouting terms like "enterprise strategy" and "adding value."

Intergraph generates envy among CAD vendors when it is gets its Imagination Engineer LT software on the Microsoft Resource Kit CD-ROM. The coup, however, has no positive effect on the fortunes of Intergraph.

Visio announces it acquired the ArchT architectural add-on from Ketiv. Later, ArchT is given away to the IntelliCAD Technical Consortium.

Several vendors, including Autodesk and Microsoft, announce a new general-purpose vector format for the Web, Vector Markup Language, VML eventually shows up in Internet Explorer v5.5 and some graphics programs, but never becomes very popular.

> **Also in June 1998**
> Graphisoft and Unigraphics Solutions go public.

July 1998

CAD vendors begin to produce year-2000 compliance statements. Under new legislation, American corporations cannot be sued for Y2K problems if they state what the problems might (or might not) be. Corporations send letters to suppliers demanding their Y2K-complience statements. Many "consultants" earn huge sums of money predicting a wild range of problems sure to occur when computer clocks and software move from 31 Dec 1999 to 1 Jan 2000; a second, smaller group of consultants says nothing serious will occur. Three groups of computer users are, however, off the hook for 38 years.

Unix, Macintosh, and Palm computers are unaffected by Y2K problems but face an absolute deadline in 2038. These two computer systems use a long integer that counts the number of seconds from January 1, 1904.

ConnectPress announces *Visio Design Solutions*, the first independent paper-based magazine for Visio Technical and IntelliCAD users. The magazine joins Visio's own quarterly *SmartSolutions* magazine and CADinfo.net's *Design Drawing*, a monthly Webzine. Later, I launch *Visions.eZine* as a biweekly e-newsletter. With the exception of *SmartSolutions*, all publications fail after a year or so. To cut costs, Visio later changes *SmartSolutions* from a print publication to a combination e-mail and Webzine.

> **Also in July 1998**
> Broderbund Software acquires Autodesk's Picture This Kitchen software.
>
> The stock market is brutal on some CAD vendors; PTC's stock value falls from $36 to $16 in less than a week because revenues are lower than expected for the first time in 40 quarters (ten years).

August 1998

Autodesk pays $520 million for Discrete Logic. The Canadian company's software creates 3D effects for movies. Autodesk merges Discrete with its Kinetix division, renaming the division New Media Group. Later, the division is renamed discrete and all product names are changed to all lowercase, such as 3d studio max — much to the puzzlement of editors (maybe it's the unix influence).

September 1998

Autodesk announces it has shipped its two millionth copy of AutoCAD. AutoCAD Release 14 shipped 750,000 copies in 15 months (works out to an average of 50,000 per month). In contrast, Visio has shipped just 12,000 copies of IntelliCAD in three months (4,000 per month).

Autodesk temporarily stops shipping Mechanical Desktop 3 because an "unprofessional message" involving the word "idiot" is found in the on-line help system. (Which reminds this book's technical editor that tech support people sometimes refer to the cause of the problem being the "ID 10t".)

IMSI says it has shipped 1.2 million copies of TurboCAD, and now has a lineup of 40 software products. Later, facing huge financial losses, the company cuts back its lineup to just a few titles, including TurboCAD.

Unigraphics Solutions ships ParaSolid 10 with 950 enhancements. The company also unveils Unigraphics 15 CAD/CAM/CAE software.

Italian CAD vendor Cad.lab, maker of Eureka Gold, hires Greg Robinson as vice president of strategic technology in charge of bringing "3D to everyone." Later, Cad.lab changes its name to think3.

Magazine editor Martyn Day starts to wonder who will benefit from object-oriented CAD. A few years earlier, it seemed that every CAD package was object-oriented, or was going to become so. Then the difficulties and drawbacks began to emerge, particularly in the area of exchanging drawings between CAD packages. Over the next year, the CAD industry would focus its marketing efforts on being all-things Internet.

Also in August 1998

With increased competition from ParaSolid, Spatial Technology is now shipping ACIS updates every few months.

Progressive Software acquires XCAD from Xitron Technologies.

Also in September 1998

PTC ships Pro/Engineer 20.

Dassault Systemes ships CATIA 5.

Evan Yares is hired as the executive director of the OpenDWG Alliance.

Florian Lauterback is the editor of the Swedish CAD journal *CAD & Rit-Nytt*.

Bristol Technology sues Microsoft for injuries through "predatory manipulation of the access to the Windows NT programming interface"; years later, Bristol eventually wins the suit.

October 1998

Autodesk ships AutoCAD LT 98, saying it has now sold 800,000 copies of its low-cost software. Autodesk's "100% pure DWG" marketing campaign work, convincing customers to purchase LT instead of the lower priced, more capable, but less-compatible IntelliCAD. To counter the campaign, Visio Corp. announces a $149 competitive upgrade to IntelliCAD from AutoCAD, AutoCAD LT, and TurboCAD.

Intergraph ships its SelectCAD v7.2 civil engineering software. The software features an identical user interface, whether running in AutoCAD, MicroStation, or IntelliCAD.

Autodesk starts shipping Architectural Desktop. Within a week, the company must warn customers of a file-damaging incompatibility with AutoCAD Map.

November 1998

A consortium of European automotive manufacturers creates DXF-II/CDS. The file format is based on DXF (developed by Autodesk) and adds extensions to make it easier for the companies to exchange data about automotive CAD designs.

SolidWorks releases a feature-based converter that reads Pro/Engineer files. This is the first converter that recognizes features, and was developed by Geometric Software Services. Later, GSS's translation technology is acquired by SolidWork's co-subsidiary, the Spatial division of Dassault Systemes.

Interest starts to grow in Linux as an alternative operating system to Windows. Although there are a small number of Linux CAD packages (including MicroStation), none creates a significant impact because the CAD industry concentrates on Microsoft Windows and NT. Computer-aided design is more than simply operating a CAD package; it also involves a host of ancillary software tools, which are in limited availability for Linux. One Linux CAD package claims to be AutoCAD compatible, but users report the product is too rough for production use. An

---

Also in October 1998

Graebert Systems ships the AutoCAD-compatible FelixCAD v4 ($595).

Diehl Graphsoft changes its MiniCAD software to VectorWorks.

PTC changes the name of the DesignWave software (acquired from its purchase of Computervision) to Pro Desktop.

Microsoft pulls its Office 97 Service Release 2 due to the many bugs.

Symantec acquires Quarterdeck for $65 million.

Also in November 1998

Jupiter Communications reports that 42 percent of top-ranked Web sites take longer than five days to respond to a customer's email, or never respond at all.

AOL buys Netscape for $4 billion in shares.

German magazine publisher IWT Magazin Verlag launches *CAD World* magazine.

IMSI lays off several vice presidents, the chief technology officer, and 20 other senior managers in an attempt to cut costs; later, a number of the employees return to IMSI.

attempt to make IntelliCAD run on Linux fails. Still, over the next few years, bits and pieces of CAD technology are released for Linux, such as OpenCascade, ACIS, and HOOPS.

December 1998

During the fall, Cad.lab had launched a $50,000 give-us-a-new-name contest at its Web site. The contest generates 40,000 entries; the winner contributes 167 entries. The winning name is think3, which is owned by a Web development company in Canada. It takes many months for think3 to obtain the www.think3.com domain name from the Canadian company. Think3 goes on to innovate by creating a 3D interactive game, called "The Monkeywrench Conspiracy," that teaches how to use their thinkdesign software by playing the game.

Also in December 1998
IMSI says Visual CADD 3.0 will be ready to ship soon.

Visio Corp. starts shipping IntelliCAD as an OEM CAD engine.

Syquest Technologies, maker of a popular line of removable drives, goes bankrupt.

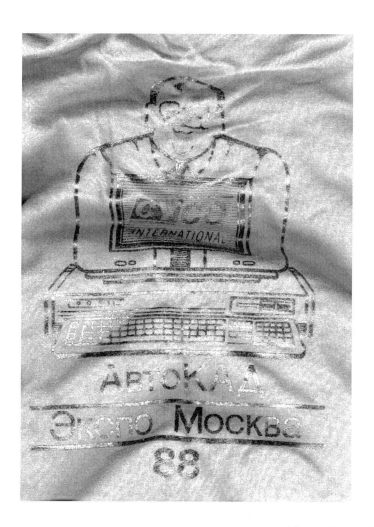

Occasionally, a T-shirt is historic. This one from Cyco commemorates the first International AutoCAD Expo in Moscow, 1988.

1999

With 1998 being such a tumultuous year, it was hard to believe that 1999 would be even more so. CAD prices would tumble to as low as $0. And the buzzword of the year is "Y2K."

January 1999

Lockheed Martin notifies its CalComp unit that it will no longer extend credit beyond the $43 million already handed out. CalComp expects to run out of money within a month. Lockheed had tried unsuccessfully to sell CalComp several times earlier, and has now decided to shut down the money-losing unit. CalComp (California Computer) was incorporated in 1958 and invented the first drum plotter. It introduced the legendary 1040 pen plotter series in 1984. Over the years, CalComp sold digitizing tablets, graphics boards, scanners, plotters, and color printers. The company had acquired Summagraphics, the inventor of the digitizing tablet, and for a while sold the Israeli-developed IsiCAD software, later renamed Cadvance.

Another long-time computer name goes bankrupt. Hayes was the inventor of the modem for personal computers. The AT command set, used by all modems, had been created by Hayes.

HP creates the Apollo Consumer Products division to brand its under-$100 printers. The irony is that when Apollo was an independent company, it made high-end workstations that competed successfully with Sun Microsystems. After HP bought the company for its workstation technology, it buried the Apollo name for nearly a decade.

SolidWorks gets a patent for "a graphical browser for computer models," the Feature Tree. This user interface element is

Above: This button, given away by CalComp at trade shows, featured three flashing LED lights.

At right: This fridge magnet, in the shape of four puzzle pieces, was given away at CAD shows by Autodesk Australia.

Also in January 1999

McNeal-Schwendler acquires Knowledge Revolution for $19 million.

Dassault Systemes acquires Euclid Styler from Matra Datavision.

---

commonly used by most mechanical CAD software packages. The company reassures competitors, however, that it has no intention of enforcing the patent.

PTC ships Windchill v2, its product development management software. Later, PTC concentrates more on Windchill sales than Pro/Engineer, to the detriment of the company. Not until early 2001 does PTC realized it needs to return to emphasizing sales of Pro/E.

Competitors continue to gun for Autodesk. For example, SoftTech attempts to lure AutoCAD users by offering its CADRA software at a 45 percent discount. "It is our goal to make CADRA the standard design, drafting, and documentation product on the market." The effort fails. In another example, SolidWorks runs three-page ads telling AutoCAD users, "Unlike Mechanical Desktop, which was built on a 15-year-old 2D foundation, SolidWorks software was designed from the ground up as a 3D tool." While SolidWorks (first released November 1995) is a highly successful product, in the years to come it will face competition from Autodesk's Inventor.

February 1999

Autodesk announces that AutoCAD 2000 will ship after a brief four-week beta period. Autodesk says the new release has 400 added and improved features, which sparks a marketing war between CAD vendors as to whose new release can boast the largest number of new features. AutoCAD's list price remains at $3,750, even though the street price is typically $1,000 less.

Autodesk also announces that a new line of viewing-markup software will be called Volo. The base product will remain free, and later is included with AutoCAD 2002.

> **Also in March 1999**
> SoftSource begins work on version 2 of the SVF (simple vector format) specification.
> PTC ships Pro/Engineer 2000i, and acquires Auxillium for $79 million.
> Iomega sells its 2GB Ditto tape products to Tecmar for $3 million.

March, 1999

Two former Autodesk dealers bring a lawsuit against Autodesk for fraud and breach of contract. Over the next years, other dealers will bring suits against Autodesk. By 2001, Autodesk announces it wants just 40 percent of its software sold by dealers, down from a onetime 97 percent; the other 60 percent is to be sold direct to customers by Autodesk itself.

VDS announces that it will employ both ACIS and ParaSolid in its IronCAD software. It is the first company to implement a dual-kernel strategy. The move foreshadows the eventual cooperation between Unigraphics and Spatial to coordinate their ParaSolid and ACIS development efforts.

April, 1999

VersaCAD returns to its founders. Tom and Mike Lazear had founded VersaCAD Corp as one of the very first PC CAD software companies in the early 1980s. They sold VersaCAD to Prime Computer, which was then acquired by Computervision, which in turn was acquired by Parametric Technology Corp. Under the three different owners, the product languished in its DOS state. PTC didn't want VersaCAD, and handed it back to the original developers. The Lazears worked hard to make contact with the remaining VersaCAD users (estimated at about one thousand), and port VersaCAD to Windows and Macintosh.

> **Also in April 1999**
>
> Visio closes the San Diego office where IntelliCAD 98 had been developed.
>
> Bentley Systems and Intergraph settle a lawsuit over the payment of royalties when Intergraph agrees to pay Bentley $15.5 million; as a result, Intergraph's stake in Bentley is reduced from 50 to 33 percent.

May 1999

Former Autodesk cto CTO John Lynch launches Design Variations. His new software uses just six commands to create 3D building designs. Most other user interface elements involve grips, snaps, and automatically generated portions. The product never becomes a retail product, but is re-architected for Web use.

Belgium-based Brics changes its name to Bricsnet. The company had developed the core technology used by MicroStation TriForma. Bricsnet releases Bricsnet Architect, which is based on IntelliCAD and the ACIS solid modeler. A number of Bentley executives make their way to Bricsnet.

June 1999

Autodesk president and chief operating office Eric Herr resigns after seven years. At the AEC Systems show, Autodesk hints at something new called Projectpoint.com, but gives no other details. For many months, the www.projectpoint.com Web site sports a single, mysterious sentence: "Through this portal

> **Also in May 1999**
>
> Intergraph ships SmartSketch 3.0, the new name for its Imagineer product.
>
> For a short time, FIT makes its Cadvance 6.5 available for $10.
>
> IMSI allows free downloads of its TurboCAD software.
>
> The Autodesk share price suffers when users fail to upgrade to AutoCAD 2000 in large numbers because so many customers are satisfied with Release 14; profits fall by 73 percent.

> **Also in June 1999**
>
> Spatial announces a version of ACIS for Linux.
>
> Unigraphics Solutions unveils Solid Edge 7.
>
> Graphisoft discontinues its pay-per-use plan for ArchiCAD, an early failure for the subscription revenue model that many software companies hope to use in the coming years.
>
> Dan Raker sells his book publishing operation, OnWord Press, to International Thompson Publishing.
>
> IMSI restructures itself, cutting three-quarters of its software lineup and one-third of its staff.
>
> Network vendors start to come out with wireless network cards for notebook computers; later, this chapter is written on a notebook computer sporting a wireless network card.

pass the finest fish in the world." Later, Buzzsaw.com CEO Carl Bass explains he saw the words over the entrance to the San Francisco fish market located across the street from the coffee shop when he and others planned the launch of their company.

July 1999

More companies report worsening revenues. Autodesk reports a net loss of $17.1 million, and a month later lays off 350 employees. PTC and Visio warn their Q3 earnings will be down. Spatial Technologies warns its Q2 revenues will fall short of estimates. Intergraph continues to lose money (losses now total $420 million), and lays off another 200 employees. Other CAD vendors, however, are doing well. Giant Unigraphics Solutions and tiny Diehl Graphsoft report increased revenues, with the latter increasing its workforce by 40 percent.

Autodesk announces that one million copies of LT have shipped. ManuSoft releases LTX, which allows AutoCAD LT to run add-ons written in ObjectARX. This results in LT gaining features found previously only in AutoCAD. LT3D from Drcauto gives AutoCAD LT 3D surfaces and solids modeling.

August 1999

Visio Corp. announces it will grant a royalty-free, nonexclusive, perpetual license for the IntelliCAD 2000 source code to a new organization it formed, the IntelliCAD Technical Consortium — after development of IntelliCAD 2000 is completed in a few months time. The purpose of the ITC is to distribute the source code, collect enhancements, and make bug fixes. According to a Visio spokesman, just 30,000 copies of IntelliCAD 98 had been sold — a number far too small for a company more familiar with selling millions of copies of software.

> **Also in August 1999**
>
> In an attempt to increase sales, Ditek reduces the price for its 2D design-drafting software to $49 and its 3D software to $199.
>
> Informatix Software ships MicroGDS v6.
>
> PTC ships Windchill 3.
>
> Unigraphics purchases Applicon.
>
> Spatial Technologies unveils 3Dmodelserver.com, which it later renames 3Dshare.com.
>
> Baystate Technologies changes its name to CADKEY, and says it has shipped 265,000 copies of its Cadkey software.

> **THE Y38 PROBLEM**
>
> Unix has a year-2038 problem because of the way it calculates dates. $2038 = 1970 + (2\wedge31)/60/60/24/365$.
>
> When the Unix clock switches to 64-bit numbers, this will become a year 292,471,210,647 problem.
>
> PalmOS computers face their year limit on December 31, 2031.

Bentley Systems is the first to propose a CAD-specific dialect for XML (extended markup language) called aecXML. Few people at the time understand the significance of the announcement. But within a year, Microsoft will announce that most of its applications and operating systems will be XML-compatible, and CAD vendors will start to add XML export to their software. Overcoming its initial reluctance, Bentley confirms it has joined the OpenDWG Alliance.

To make an aggressive push into 2D mechanical CAD, Unigraphics Solutions announces it will give away 500,000 copies of Solid Edge Origin, a 2D subset of Solid Edge. The 3D version, called Solid Edge Origin 3D, is sold at $495.

Autodesk ships AutoCAD LT 2000, and raises its price to $559, moving it out of the under-$500 category. Visio ships Visio 2000 Standard Edition, followed by the Technical, Professional, and Enterprise editions. Autodesk counters by shipping Actrix Business for just $99.

September 1999

Visio Corp. announces a plan to sell itself to Microsoft for $1.3 billion, Microsoft's largest purchase ever. There are two reasons for the sale: (1) boost the Visio share price; and (2) gain access to Microsoft's international salesforce. Microsoft grandly calls Visio "the third leg" of a stool that consists of words (Word), numbers (Excel), and diagrams (Visio). The speculation that Microsoft will bury Visio Standard in its Office suite turns out to be false; Microsoft does, however, kill off Visio Technical.

> **Also in September 1999**
>
> Unigraphics announces UGS 16.
>
> Sun buys StarOffice, then gives it away free, and announces that a browser-based version is planned. Microsoft immediately announces it will copy the move by Webifying its Office software.
>
> 3Com announces it will sell its Palm division. Handspring starts shipping the PalmOS-based Visor handheld computer.

Meanwhile, Visio's planned giveaway of the IntelliCAD 2000 source code runs into a problem when people realize that many crucial components — accurate plotting, rendering, VBA programming, raster display, DWG read-write, etc. — are based on code licensed from other companies, and cannot be given away. Later, the IntelliCAD Technical Consortium negotiates individual licenses for its members.

Autodesk unveils Inventor, a 3D mechanical package based on all-new code. The company plans an aggressive upgrade schedule, with a new release every three or four months.

Think3 begins selling it software on-line at an annual subscription of $1,995. As an incentive, the company offers its salesforce shares in Think3.

Hercules Computer Technology, a pioneer in the graphics board industry, files for bankruptcy. Later, a Canadian company, Guillemot, acquires the 17-year-old Hercules for $1.5 million.

October, 1999

SolidWorks creates a new application that e-mails drawings with an integrated viewer. The free eDrawings product compresses

SolidWorks, AutoCAD, or DXF files. The integrated viewer allows regular viewing, 3D shaded renderings, and animations.

The same week that the IntelliCAD Technical Consortium opens its Web site, Surya Sarda opens his own download-IntelliCAD-for-free Web site at www.cadopia.com. Mr. Sarda was part of the original IntelliCAD development team.

Unigraphics publicly demos its CAD software running on Intel's 64-bit Itanium CPU. Delays, however, dog this "next generation" CPU. Systems using the Itanium are not generally available for another two years.

November 1999

Joel Orr starts tracking extranets via his *Extranet News* e-newsletter and Web site. The term "extranet" refers to using the Internet to access services external to your firm, such as project management, over a secure link. More recently, extranets have also been known as "AEC dot.coms" and "ASPs" (application service providers).

Autodesk's first dot.com venture, Buzzsaw.com, goes live. This is the Web site that Autodesk mysteriously referred to at AEC Systems earlier in the year. ProjectPoint is the name of the project management software that runs at Buzzsaw.com. Later, Autodesk launches additional Web sites, including Point A (a portal for its software users), RedSpark.com (a manufacturing exchange), and Streamline (design management for mechanical CAD).

The United States Justice Department starts investigating Microsoft's acquisition of Visio Corp. The investigation delays the acquisition by a month, but the department allows the purchase to go ahead. Visio announces its revenues in fiscal year 1999 were $200 million, up 20 percent from the previous year.

Autodesk spins off the Pro Landscape business unit to a new company called drafix.com. Its original owner, Augie Grasis, heads up the "new" company. The company was originally known as Foresight Resources, and its CAD software had been called Drafix until acquired first by Softdesk, then by Autodesk.

December 1999

Alibre reveals itself as a subscription-based mechanical CAD design service. The company is partially funded and headed up by J. Paul Grayson, the former CEO of Micrografx, best known for its diagramming software and its annual Chili for Children Cook-Off at Comdex.

CEO Michael Cowpland says that within five years half of Corel's revenue will come from Linux. Less than a year later, Mr. Cowpland is ousted from the company, and Corel is looking to unload its unsuccessful Corel Linux operating system.

Also in October 1999

AutoCAD ships AutoCAD OEM 2000 to restricted licensees, as well as AutoSketch Release 7.

IronCAD 3.0 is available for $4,995.

Intel starts shipping the Pentium III at speeds starting at 700 MHz.

Think3 ships Thinkdesign 4.

Intergraph releases SelectCAD v8.

Also in November 1999

CADKEY 99 starts shipping. Bricsnet launches its www.Bricsnet.com Web site.

2000

Within minutes of computer clocks changing from 23:59:59 99-12-31 to 00:00:01 00-01-01, the phrase "Y2K" becomes passé. The buzzword of the year is "dot.com."

January 2000

Microsoft completes its purchase of Visio for $1.4 billion. But Visio employees discover working under Microsoft is not at all what they had imagined. A number of vice presidents leave when they find out they won't be keeping the prestigious title. Parts of the Visio code have to be ripped out and replaced with Microsoft-approved components. Press releases become rare because they must fit in with the Microsoft schedule. As for the original reasons for selling the company: (1) within a few months, Microsoft shares fall to half their value; and (2) the international sales force sees Visio as just one more product to hawk.

February 2000

IMSI finally confronts its financial problems. It reduces its workforce to fifty; closes international offices; removes its listing from the NASDAQ stock exchange; and tries to sell off most of its software. The trigger is an arbitration decision that goes against IMSI, costing the company $2.2 million. IMSI's total debt is $42 million, but its market value is just $1 million. The company's CEO says he will transform the company into "a great Internet competitor."

Another under-$500 CAD vendor, Ditek, morphs itself into HomeProject.Com. The Canadian company is better known for its DynaCAD software. A year later, the Web site appears to be going strong.

Bricsnet announces it is the first to commercialize the now-free IntelliCAD 2000. The company will distribute and support IntelliCAD in Europe and the United States. Struc Plus signs a similar agreement for Australia, New Zealand, and South East Asia. A year later, Struc Plus gives up, citing too many people (who got IntelliCAD free by downloading it) expecting to get support and training for free.

Autodesk starts shipping Actrix 2000 Technical. Sales, however, prove poor enough that Autodesk starts giving away Actrix in boxes of AutoCAD LT 2000.

Autodesk opens its e-Store to start selling some of its software (but not AutoCAD, at least not yet) via the Web. First, though, the company must come to an agreement with its resellers on sharing profits from selling AutoCAD via the Web, which bypasses the dealer. Later, Autodesk releases briefly an ASP version of AutoCAD that is

---

Also in January 2000

Matra Datavision opens the source code to its CAS.CADE modeling object libraries, renaming it Open CASCADE.

Autodesk, Intergraph, and several others get together to define LandXML for software related to land planning, civil engineering, and surveying. www.landxml.org

The first virus unique to Visio is detected, infecting template and stencil files.

Spatial Technology ships ACIS 6.

AOL buys MapQuest.Com for $1.1 billion.

Graphisoft acquires Cymap.

able to run within your Web browser anywhere you have access to the Internet.

Autodesk Ventures purchases nine percent of CapacityWeb. A year later, CapacityWeb goes out of business. Other investments by Autodesk Ventures are more successful.

Nemetsheck finally figures out how to attack the North American market, after several failures. The company purchases Diehl Graphsoft for $30 million, and renames it Nemetschek NA (North America). Rather than trying to sell allPlan to USA, Nemetschek NA continues to sell its successful VectorWorks.

> **Also in February 2000**
> Unigraphics Solutions ports its ParaSolid modeling kernel to Linux.

March 2000

On March 10, the NASDAQ stock index reaches its all-time high of 5,049. The peak marks the end of two years of "irrational exuberance" over dot.com fever. One year later, the index will fall to 2,053. Some 420 companies listed on NASDAQ see their share price fall by 90 percent. As a business writer in *Fortune* puts it, 2000 saw the "greatest legal evaporation of wealth in history."

SolidWorks launches a restraining order against Alibre, one business day before Alibre is due to launch its Web-based mechanical design collaboration service. The order prevents Alibre from contacting any SolidWorks employees, or use competitive information the company may have obtained from SolidWorks. It appears the tussle is over a SolidWorks employee hired by Alibre, following which Alibre began recruiting additional employees from SolidWorks.

Microsoft is found guilty of monopolistic behavior. The judge recommends splitting the company in two: operating systems and applications software. Microsoft avoids the split by appealing the order.

April 2000

Microsoft hands over ArchT to the IntelliCAD Technical Consortium. Microsoft had ended up with the architectural design software as a result of its purchase of Visio Corp. Curiously, Microsoft decides to hang on to ArchT's 2D and 3D symbol libraries. Meanwhile, Matt Richard, original author of the OpenDWG Toolkit, leaves Microsoft to lead a non-CAD life.

Revit Technology Corporation launches itself and its Revit architectural software during a media splash at Harvard University in Cambridge, Massachusetts. The software uses the single building model (as do ArchiCAD, Triforma, and others) and parametrics. The founders of the company used to be with Parametric Technology Corporation. The software itself is not ready for the launch, but ships a month later.

A company tries to sell CAD-related domain names for $395,000 each, names like Computer-Automated-Design.com. Too bad the term is actually computer-*aided* design.

May 2000

MicroGDS v6.1 becomes the first CAD software to support XML — by accident, it turns out. Informatix was looking for a way to export MicroGDS drawings in ASCII format. Looking over various ASCII standards, they decide on XML. They have no idea of the impact of their decision until they began reading press release after press release of how other CAD vendors plan to support XML sometime in the future.

Microcadam stops selling Helix Design System outside of Japan. Users are encouraged to switch to CATIA v5 from Dassault Systemes.

Autodesk invites the media to New York City, where the next release of AutoCAD 2000i is demonstrated. At the same event, the company shows a secret new project — called Project Nora — that seeks to redefine architectural software completely. Named StudioDesk, the software runs on a tablet computer and mimics the desk of an architect who does not use a computer. The www.studiodesk.com Web site has a Java-based demo version, where architects are asked to provide feedback. Some industry observers feel that StudioDesk will eventually replace Architectural Desktop.

Number Nine, a prominent graphics board manufacturer, goes out of business. Other companies that have exited the graphics chip/board business include S3, Intergraph, NeoMagic, Gigapixel, and ArtX.

> **Also in May 2000**
> Bentley Systems forms its own ASP called Viecon.com.

June 2000

A/E/C Systems 2000 has its hottest show in several years. The reason? All that money being thrown around by the suddenly rich dot.coms. There are so many giveaways from the booths that I return home with a suitcase full of squeeze toys, t-shirts, pens, CDs, mouse pads, pins, buttons, and more. But the vast amount of money being spent by the new dot.coms worries me, and I title my show report "Waiting for the Bubble to Burst."

IMSI finds a new home for Visual CADD. TriTools Partners agrees to take over the development of the orphaned software, which has now been shuffled from Numera to Corel to IMSI to TriTools.

July 2000

Spatial announces that it is selling its ACIS component to Dassault Systemes of France for $21.5 million. Curiously, the news is first released by Autodesk, one day before Spatial makes its announcement. By purchasing ACIS, Dassault is in a bind, however, because its SolidWorks division uses the solid modeling kernel from archcompetitor ParaSolid. Will SolidWorks be forced to switch over? Not for now.

Autodesk ships AutoCAD 2000i, which is heavily promoted as "the Internet release." Initial sales are slow, as heads-down drafters consider the Internet not their primary interest.

August 2000

P2P (peer to peer) software starts to make an impact, led by software with names like Napster, Gnutella, and Groove. PTC says it plans to be the first CAD vendor to add Groove technology to its Windchill collaboration software. Other CAD vendors chime in that they already have P2P in their software, or else try to make it appear their CAD software is capable of P2P-like activities. Less than a year later, Microsoft will announce it intends to make its instant messaging (a.k.a. P2P) software the basis of its other software.

Autodesk announces no fewer than four XML initiatives, some in conjunction with other CAD vendors: adpML for electronic design catalogs; DesignXML for CAD; LandXML for land development; and RedlineXML for marking up drawings.

September, 2000

A competing offer from SDRC forces Dassault to raise its offer for Spatial to $25 million. The sale is completed in November. SEC documents filed by Spatial shows that SRDC was the first to secretly propose to buy Spatial, but Spatial prefers Dassault's all-cash offer. Licensees of ACIS are worried that Dassault will cut

Also in September 2000
Nemetschek NA says that the number of VectorWorks seats have reached 105,000.

Upperspace Corp releases DesignCAD 3000 with 2D drafting and 3D modeling for $300.

Autodesk releases a test version of Actrix 2000 Technical as an ASP product.

Also in October 2000

UGS ships Solid Edge v9.

SolidWorks sells its 100,000th copy of SolidWorks.

Autodesk announces that new features for AutoCAD and vertical products (but not LT) will be released as "Extensions" every three months or so; the price for the Extension modules starts at $99.

Visual CADD v4 becomes available from www.visualcadd.com hosted by CADDvillage.

PTC's share price jumps upon takeover rumors, which prove false.

Intel plans to start shipping the 1.4 GHz Pentium 4 CPU in November, and hopes to push the speed to beyond 2 GHz by mid-2001.

Iomega changes the name of its 40 MB Clik drive to PocketZip due to lackluster sales; later the capacity is increased to 100 MB but Iomega adds "digital content protection."

Desktop computer sales slow while handheld computer sales nearly triple.

Microsoft purchases a quarter of Corel for $135 million, but reconsiders the decision when the FTC begins to investigate.

---

them off, but Dassault promises to honor all ACIS contracts. Industry observers are surprised that Autodesk doesn't purchase ACIS, since it has the largest number of licenses, but Autodesk says they have no interest getting into the component business. The non-ACIS portion of Spatial changes its name to PlanetCAD. Archrival Unigraphics counters with a make-no-payments-for-one-full-year offer to licenses of ParaSolid.

In a case of history repeating itself, GraphStore announces a software package that can edit MicroStation 2D DGN design files. (Recall that Bentley got its start with MicroStation, whose purpose was to edit Intergraph files.) Later the same month, Pangaea CAD Solutions announces it plans to release DualCAD, a program that edits both AutoCAD and MicroStation drawings.

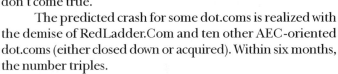

Bentley begins to release information about MicroStation V8. The most significant change is the DGN2 file format, which gets a massive overhaul to bring it up-to-date with the 1990s and make it more compatible with AutoCAD. The company switches top executive positions: Keith Bentley becomes chief technical officer, Greg Bentley becomes chief executive officer, while marketing vp Yoav Etiel leaves the company, ending up at Bricsnet. Long-time rumors that Bentley will go public continue to swirl, but don't come true.

The predicted crash for some dot.coms is realized with the demise of RedLadder.Com and ten other AEC-oriented dot.coms (either closed down or acquired). Within six months, the number triples.

October 2000

Motiva abruptly ceases operations on Friday the 13th. Employees, customers, dealers, and even investor Autodesk are taken by surprise. The power of the Internet allows everyone to learn of the shutdown within one business day, unlike a decade or two earlier when the news might have taken months to spread via print magazines. Ex-employees form a new company called exMotiva, and hope to acquire the source code, while competitors rush in to lure orphaned customers. One industry observer wonders whether the rise of peer-to-peer Groove software and the fall of client-server Motiva is a coincidence.

Also in November 2000

Eagle Point CEO Rod Blum leaves the company.

CADSOFT is acquired by Master Builder Design Concept.

Seagate ships an 180 GB hard drive for $2,195, the price of a 20 MB hard drive twenty years earlier.

CNET reports the 130 Internet companies have closed their doors since the beginning of 2000.

Autodesk says it wants to aggressively move into subscriptions.

SolidWorks acquires CIMlogic, and demonstrates an innovative user interface that eliminates the need for dialog boxes.

Squeeze toys became a popular tradshow giveaway in the new millennium.

November 2000

Graebert Systems, maker of the AutoCAD-compatible FelixCAD, ports the software to the PocketPC environment. The company says it has shipped 50,000 copies of the desktop version of the software.

After visiting Comdex, Jon Peddie Associates comes to the conclusion that: (1) formats don't matter; (2) platforms don't matter; but that (3) the network is everything.

December 2000

Cadkey reverts to the original numbering system. The previous release was called Cadkey 99; the next release would be called Cadkey 19. Cadkey says Release 20 will be an all-new CAD architecture that combines 2D with 3D wireframe, surfaces, solid modeling, and parametrics — the result of four years of development.

Intergraph launches a new division called IntelliWhere for location-based services (LBS). This controversial technology allows data to be sent to a wireless devices based on its location. LBS combines mobile phone technology with GIS (geographic information systems). A month later, IBM and Autodesk unveil their LBS divisions.

Also in December 2000

Bruce Morgan resigns as CEO of PlanetCAD.

Voodoo graphics board manufacturer 3dfx sells its assets to competitor nVidia.

2001

The real start of the new millennium. The buzzword of the year is "P2P" (peer to peer).

January 2001

CAD companies continue on acquisition binges, and try to change their revenue model from perpetual licenses to subscriptions.

Bentley completes its acquisition of Intergraph's civil engineering, plot services, and raster conversion software for $40 million.

Bricsnet reports it has acquired nine companies in the last year.

PTC rents its software by the month or by the quarter (three months) at prices that start at $874 for thirty days.

Also in January 2001
IMSI ships TurboCAD v7.
Alventive ships IronCAD 4.
Archway Systems ships VersaCAD Mac 2000.
Toshiba fits 2GB hard drive on a PC Card for $600.
DataPlay has a 1" write-once disk holding 500MB.

February 2001

Delcam of England boasts the first direct translator for Pro/Engineer 2000i2 files. The event is significant because PTC had encrypted the file format. *CADdesk AEC* editor Martyn Day describes how Delcam accomplished the task: "To access the information within the files, Delcam put a team of Russian developers (industry folklore says that these guys were all ex-KGB hackers) to work out the key to the encryption algorithm used in Pro/E 2000i. Once they had cracked that, they had to work out the key for the harder 2000i2 format. It took a year."

The Nemetschek share price spikes when a German newspaper speculates the company is prepping for a takeover. The company denies the rumor.

The Spatial Technology division (located in the United States) of Dassault Systemes (located in France) acquires the translation source code from Geometric Software Solutions (located in India). GSS has been one of the most prolific companies at writing translators between CAD systems. Just like Visio Corp.'s purchase of MarComp, Spatial promised to honor all existing contracts with competitors.

Autodesk announces that there will be an "aggregation release" of AutoCAD that consists of all previous extensions released for AutoCAD 2000i. The company's CEO says she wants words like "big-R" and "upgrade" to go away. The big-R upgrade will be known as AutoCAD 2002. Autodesk's 3D Studio MAX is used to create *Myst III: Exile*. With Release 4, Autodesk changes the name of the software to 3ds max ($3,495) — all lowercase and missing "Studio."

*In February, Unigraphics Solutions changes its name to **UGS** with a new logo that "... highlights the multiplying effect of its Solution and Platform technologies... ".*

*Other companies change their names: Varimetric to **VX**. Hammon, Jensen, Wallen & Assoc to **HJW GeoSpatial**. Composite Design Technologies to **Vistagy**. Advanced Numerical Methods to **Idelix**.*

Drcauto ships AccuRender 3 modified for AutoCAD LT, allowing the software for the first time to render 3D models with raytracing and radiosity.

The Internet is being used by 56 percent of the adult population in the United States.

March 2001

In a stunning move, UGS and Dassault's Spatial Technology division agree to cooperate on their ParaSolid and ACIS kernels by exchanging licenses for their solid modeling technologies. They hope to improve the translation of 3D drawings between CAD applications. For years, the two companies had fought hard against each other with frequent product updates (usually outstripping the ability of CAD vendors to implement to the latest version), and a form of price war.

In an attempt to boost its share price, Eagle Point Software agrees to sell itself to one of its board members. The agreement calls for former CEO Rod Blum to purchase the money-losing Bulding Design and Construction and Structural divisions.

PlanetCAD backs off its hard push into Web-based services after a poll of its customers reveals that most were not ready to do business on the Web. The company converts its 3DShare Web-based software into enterprise software, and delays further work on its Web services for a half-year.

Autodesk and Revit Technology launch a public relation war with each other over the results from the *CAD Software Evaluation Report* published by CAD for Principals. Controversy erupts when Revit announces: (1) it sponsored the report; and (2) it won the competition over five other architectural CAD packages. Autodesk complains its software was not fairly evaluated, while study participants insist Revit made no attempt to influence their findings.

Parametric Technology announces its new Granite One initiative, which opens its Pro/Engineer software for developers. This allows third-party applications to work directly with Pro/E files, without losing model associations.

Bentley Systems releases beta 1 of MicroStation V8 to its 200,000 software subscribers, while Microsoft starts selling beta copies of Visio 2002 for $8.

Bricsnet ships IntelliCAD 2000 version 2.2 ($199) with more features that take it further along the road to compatibility with AutoCAD.

Insight Development ships a beta of Squiggle v4 ($99), the software that makes DWG, DXF, and HPGL/2 plot files look like

This Delft ceramic shoe was handmade in Holland for Cyco Automation, who gave the shoe away to its dealers and to members of the CAD media.

hand-drawn art. (I launched this product's career by accident. The developer, Premisys, gave me a diskette during party at an AEC Systems show. The diskette was meant only to contain an example of their programming ability; I misunderstood, reviewed it in *CADalyst* magazine, and a puzzled Premisys began getting calls from people wanting to buy Squiggle. Premisys later sold the product to Insight.)

Alventive spins off its IronCAD software to an independent company called — no surprise here — IronCAD LLC. Alventive plans to concentrate on on-line collaborative design.

April 2001

GIS software (geographic information system) vendor ESRI ends sales and support for ArcCAD (CAD-based mapping) and Atlas GIS (business mapping). ArcCAD has been built to work with AutoCAD Release 11 in 1992. The unprecedented collaboration of ESRI and Autodesk surprised many back then. ESRI survived AutoCAD's move to Windows, the unlucky Release 13, and even AutoCAD Map (Autodesk's competing entry into the CAD/GIS space). After the launch of AutoCAD Map, however, Autodesk decided that ESRI was no longer welcome in its developer program, and ESRI had a tough time keeping up with Autodesk's pace without support.

GIS works best in its own software environment. All attempts to merge GIS with CAD have not done well, from Dennis Klein's FMS/AC (gone), to Bill McKenzie's Geo/SQL (still around), to Intergraph's MGE (overshadowed by GeoMedia), to ESRI's ArcCAD (discontinued), to Autodesk's AutoCAD Map and Land Development Desktop (four times as many GIS users employ plain AutoCAD) to Bentley's MicroStation Geographics (originally Mizar's MicroGIS). Try as they might, splicing GIS genes into CAD software does not make a graceful creature.

The CPUs embedded in these keychains from Intel are real but nonfunctioning chips. They failed the testing process.

In the future, look for other solutions to solve the CAD/GIS integration challenge, such as Windows-based desktop GIS (Intergraph's GeoMedia, for example), and client-server solutions, like Autodesk's own GIS Design Server.

— *Adena Schutzberg, editor GIS Monitor*
www.tenlinks.com/mapgis

Free downloads (from sites such as www.cadopia.com) of IntelliCAD 2000 number 200,000. Web4Engineers creates an ASP version of IntelliCAD 2000, which allows registered users (at www.web4engineers.com) to run IntelliCAD in a Web browser, store its DWG files on Web4Engineer's online filing system, and markup the drawings using eReview, also from Web4Engineer.

IBM announces it will ship in May a 20.8" LCD monitor ($5,929) with a resolution of 2048 x 1536, along with S-Video and RCA connectors for direct TV and VCR input. NEC debuts a prototype of a 61" plasma display with 1365 x 768 resolution and a 700:1 contrast ratio. The price of 15" LCD monitors falls to $500.

TriTools Partners takes over all rights to develop and market Visual CADD independent of IMSI, while IMSI retains the right to continue to sell version 4. All future versions are marketed under the TriTools Partners name.

Three of the big names in high-end CAD — Parametric Technology, SDRC, and UGS — feel the slowdown of the economy, particularly in the United States. They report that it is taking longer for larger orders to be approved: an extra signature is required here, an extra month is added to the approval process there.

Gartner reports that Dell Computer became the largest computer manufacturer in early 2001. Its 12.8 percent share of the worldwide market beats Compaq, which had been #1 for seven years. Personal computer shipments in the United States declined by 3.5% in 2000.

European regulators announce they will probe into "abusive marketing practices" by Intel, while expanding their probe of Microsoft.

"Whatever is, has already been; and what will be, has been before."
  — *Ecclesiastes*

Pins from (top to bottom): Man+Machine, The MicroStation Community, Autodesk, and CADPIPE. Below, the AutoCAD R12 pin.

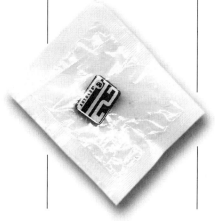

Resources for CAD Managers

STARTING POINT

The following Web sites and publications provide you with information on CAD software, hardware, and management:

Tenlinks.Com
Links to almost all other CAD-related Web sites
www.tenlinks.com/CAD/products/index.htm

upFront.eZine
Free weekly email newsletter carrying news about CAD, published by author Ralph Grabowski.
www.upfrontezine.com

The CAD Depot
A large collection of shareware and freeware for CAD users.
www.caddepot.com

CADinfo.net
A Webzine that concentrates on articles and reviews of CAD software and hardware.
www.cadinfo.net

Multi-CADD
Australian independent magazine for all CAD products.
www.echomags.com.au

APPENDIX

Here is a listing of Web sites that provide the CAD manager with more information, including CAD vendors and standards bodies.

CADserver
The Web site from the British publishers of *CADdesk AEC* and *MCAD* magazines
www.cadserver.co.uk

Cadence Channel
The Web site from the publisher of *Cadence* magazine.
www.cadenceweb.com

CAD Online
The Web site from the publisher of *CADalyst* magazine.
www.cadonline.com

MCAD Vision
Mechanical CAD articles, reviews, and opinion by MCAD Cafe.
www.mcadcafe.com/MCADVision

Desktop Engineering
www.deskeng.com

DesignNews Online
www.manufacturing.net/magazine/dn

CAEnet
The Web site from the publisher of *Computer-Aided Engineering* magazine.
www.caenet.com/res/index.html

Pro/E - The Magazine
www.proe.com

Solid Solutions magazine
www.solidmag.com

CATIA Solutions magazine
www.catiasolutions.com

The early packaging for AutoSketch (Autodesk) was a small, bright yellow, spiral-bound binder.

CAD VENDORS

The following list of CAD vendors is, of necessity, incomplete and may be out-of-date due to mergers, closures, and changes. For example, as this book was being written, Varimetrix changed its name to VX, and its Web address to www.vx.com (the old address, www.varimetrix.com, still works for the time being).

The following listing segregates CAD software products by operating system: Windows, Macintosh, and Linux.

CAD for Windows

Alibre
Alibre
www.alibre.com

allPlan
Nemetschek
www.nemetschek.com/en/products/allplan

ArchiCAD
Graphisoft
www.graphisoft.com

ArchiTECH.PC
SoftCAD International
www.softcad.com

ArchT
Eagle Point Software
www.eaglepoint.com

ARRIS
Sigma Design International
www.arriscad.com

AutoCAD
Autodesk
www.autodesk.com/autocad
AutoSketch
www.autodesk.com/products/asketch

Corel always gives its products colorful packaging, such as Corel CAD illustrated here.

Bricsnet Architecturals
Bricsnet
www.bricsnet.com/about/services/architecturals

Cadkey
Cadkey
www.cadkey.com

Cadsoft Build
CADSOFT
www.cadsoft.com

CADStd
Apperson and Daughters
www.cadstd.com

Cadvance
FIT
www.cadvance.com

CATIA
Dassault Systemes
www.catia.com

Chief Architect
Advanced Relational Technology
www.chiefarchitect.com

DataCAD
DataCAD LLC
www.datacad.com

DesignCAD
Upperspace
www.designcad.com

DESI-III
H. Mariën
users.pandora.be/desi-iii/index.html

FastCAD
Evolution Computing
www.fastcad.com

IBM spun off CAD/3X to an independent company named Altium.

FelixCAD aka FCAD
Graebert Systems
www.fcad.com

Generic CADD
CaddVillage; discontinued by Autodesk
www.genericcadd.com

I-DEAS
SDRC
www.sdrc.com/ideas

IDRAW 2000
Design Futures
www.designfutures.com

IntelliCAD 2000
IntelliCAD Technical Consortium
www.intellicad.org and www.cadopia.com

IronCAD
IronCAD LLC
www.ironcad.com

JustCad
JustCad
www.justcad.com

ME10
CoCreate Software
www.cocreate.com

MEDUSA
PTC
www.ptc.com/products/medusa/drafting.htm

MicroGDS
Informatix
www.informatix.co.uk

Mannequin was a product of HumanCAD, a division of Biometrics Corp of America.

MicroStation
Bentley Systems
www.bentley.com/products/index.htm

MultiCad
MultiQuant
www.multi-cad-c.com

PC Draft
Microspot
www.microspot.com/software/pcdraft.html

Pro/ENGINEER
Parametric Technology Corp
www.ptc.com/products/flex_eng.htm

Project Architect
AEC DesignWare
www.aecdesignware.com

QuickCAD
Autodesk
www.autodesk.com/quickcad

Revit
Revit Technology Corp
www.revit.com

SmartSketch a.k.a. Imagineer Technical
Intergraph
www.intergraph.com/smartsketch

Solid Edge
Unigraphics Solutions
www.solid-edge.com

SolidMaster
CADMAX
www.cadmax.com

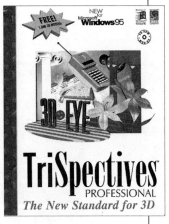

TriSpectives Professional, produced by 3D/eye, may well have been the most exciting new concept in CAD software in the 1990s.

SolidWorks
SolidWorks Corp
www.solidworks.com

SwissPrecision/Engineer
Precisionsoft AG
www.swissprecision.com

T-FLEX Parametric
Martin Sales International
www.tflex.com/products.html

thinkdesign
think3
www.think3.com/products/products_td.htm

Tlinea
Inicio
www.iespana.es/tlinea/indexi.htm

TurboCAD
IMSI
www.turbocad.com

Unigraphics
UGS
www.ugs.com

Uni-Tool
Advanced EMC Solutions
www.aemcs.com/patrucco_main.html

Vdraft aka Virtual Drafter
SoftSource
www.vdraft.com

VeCAD
Comandor
www.comandor.khv.ru/vecad.htm

The packaging for VersaCAD/386 while it was sold by Computervision.

VectorWorks
Nemetschek North America
www.nemetschek.net

Vellum Draft
Ashlar
www.ashlar.com/Products/Draft_99

Visual CADD
IMSI
www.imsisoft.com/products/visualcadd

VX Vision
VX Corp; formerly Varimetrix
www.varimetrix.com

CAD for Macintosh

ArchiCAD
Graphisoft
www.graphisoft.com

Architrion
BAGH Technologies
www.bagh.com

CADintosh
Lemkesoft
www.lemkesoft.de/us_cadabout.html

MacDraft
Microspot
www.microspot.com/software/macdraft.html

PowerCADD
Engineered Software
www.engsw.com

RealCADD
adX
www.adx-online.com/realcadd/realcaddus.htm

XCAD was the first under-$500 solid modeling CAD software package.

CAD for Linux

ARCAD
ARCAD Systemhaus
www.arcad.de

CADStd
Apperson & Daughters
www.cadstd.com

CEDRAT
(electromechanical and thermal engineering)
www.cedrat-grenoble.fr

CYCAS
(2D and 3D architectural)
www.cycas.de

FREEdraft
(2D mechanical)
freeengineer.org/Freedraft/index.html

gCAD
(2D CAD)
gaztelan.bi.ehu.es/~inigo/gcad

GIGVIZ
(3D design visualization)
www.gig.nl

ICAADS
(AEC and land development)
www.icaads.com

LinuxCAD 2000
(AutoCAD-compatible)
www.softwareforge.com

ME10 for Linux
CoCreate
www.cocreate.com

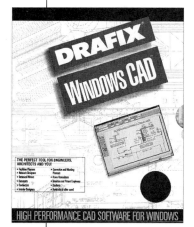

Drafix was the first CAD software released for the Windows operating system.

MicroStation 95
(academic version only) Bentley Systems
www.bentley.com/products/microstation95/

MSC.Linux
MSC Software
www.mscsoftware.com

OCTREE
(architectural drawing, modeling, and visualization)
www.octree.de

Open CASCADE
(development of 3D mechanical, AEC, and GIS tools)
www.opencascade.com

VariCAD
VariCAD
www.varicad.com

Varimetrix
(mechanical)
www.varimetrix.com

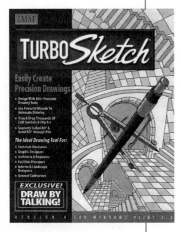

IMSI originally released TurboSketch as low-cost CAD; it was replaced by the free version of TurboCAD.

Standards Bodies

World Standards Services Network
www.wssn.net/WSSN/index.html

The Construction Specifications Institute
www.csinet.org

American Institute of Architects
www.aia.com
www.e-architect.com

USA National CAD Standard
www.nationalcadstandard.org

Construction Specifications Canada
www.csc-dcc.ca

International Organization of Standards
www.iso.ch

Most of the following international standards Web sites have an English version:

Country	Organization	Web Site
Argentina	IRAM	www.iram.com
Australia	SAIA	www.standards.com
Austria	ON	www.on-norm.at
Belgium	IBN	www.ibn.be
Bosnia & Herzegovina	BASMP	www.bih.net.ba/~zsmp
Brazil	ABNT	www.abnt.org
Canada	SCC	www.scc.ca
China	CSBTS	www.csbts.net
Colombia	ICONTEC	www.icontec.org
Costa Rica	INTECO	www.webspawner.com/users/inteco
Croatia	DZNM	www.dznm.hr
Czech Republic	CSNI	www.csni.cz
Denmark	DS	www.ds.dk
Equador	INEN	www.ecua.net/inen
Egypt	EOS	www.misrnet.idsc.gov
Finland	SFS	www.sfs.fi
France	AFNOR	www.afnor.fr

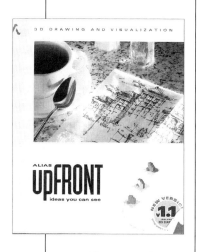

Alias Research distributed upFront as 3D drawing and visualization software in the early 1990s.

Country	Organization	Web Site
Germany	DIN	www.din.de
Greece	ELOT	www.elot.gr
Hungary	MSZT	www.mszt.hu
Iceland	STRI	www.stri.is/stri
India	BIS	wwwdel.vsnl.net.in/bis.org
Indonesia	BSN	www.bsn.go.id
Ireland	NSAI	www.nsai.ie
Israel	SII	www.iso.co.il/sii
Italy	UNI	www.unicei.it
Japan	JISC	www.hike.te.chiba-u.ac.jp/ikeda/JIS
Kenya	KEBS	www.kebs.org
Korea, Republic	KATS	www.ats.go.kr
Luxembourg	SEE	www.etat.lu/SEE
Malaysia	DSM	www.dsm.gov.my
Morocco	SNIMA	www.mcinet.gov.ma
New Zealand	NZ	www.standards.co.uk
Norway	NSF	www.standard.no/nsf
Philippines	BPS	www.dti.gov.ph/bps
Poland	PKN	www.pkn.pl
Portugal	IPQ	www.ipq.pt
Russian Federation	GOST R	www.gost.ru
Saudi Arabia	SASO	www.saso.org
Singapore	PSB	www.psb.gov
Slovenia	SMIS	www.usm.mzt.si
South Africa	SABS	www.sabs.
Spain	AENOR	www.aenor.es
Sri Lanka	SLSI	www.naresa.ac.lk/slsi
Sweden	SIS	www.sis.se
Switzerland	SNV	www.snv.ch
Thailand	TISI	www.tisi.go.th
Trinidad and Tobago	TTBS	www.opus.co.tt/ttbs
United Kingdom	BSI	www.bsi.org
United States	ANSI	www.ansi.org
Uruguay	UNIT	www.unit.org
Venezuela	TCVN	tcvn.vnn.vn

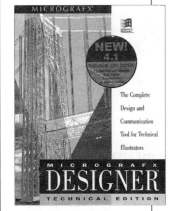

Micrografx marketed Designer Technical Edition as technical illustration software for CAD users.

Color-Pen Table

B

Some CAD systems match entity color to pen number. The United States Coast Guard Civil Engineering Technology Center assigns the following 255 combinations of pen widths and colors to MicroStation and AutoCAD drawings.

APPENDIX
This appendix provides an entity color-pen matching table.

Color Number	Pen Plotter mm	Laser in.	Plot Color	MicroStation Color No.	Linewt.	AutoCAD Color No.
1	0.18	0.007	Black	3	0	1
2	0.25	0.010	Black	4	1	2
3	0.35	0.014	Black	2	2	3
4	0.35	0.014	Black	7	2	4
5	0.50	0.020	Black	1	3	5
6	1.00	0.039	Black	5	7	6
7	1.40	0.055	Black	0	10	7
8	0.35	0.014	Halftone	9	2	8
9	2.00	0.079	Black	14	15	9
10	0.18	0.007	Black	10	0	10
11	0.25	0.010	Black	19	1	11
12	0.35	0.014	Black	27	2	12
13	0.50	0.020	Black	35	3	13
14	0.70	0.028	Black	43	5	14
15	1.00	0.039	Black	51	7	15
16	1.40	0.055	Black	59	10	16

RESOURCE

The United States Coast Guard Civil Engineering Technology Center has its Web site at www.uscg.mil/mlclant/cetc/index.htm

Color Number	Pen Plotter mm	Laser in.	Plot Color	MicroStation Color No.	MicroStation Linewt.	AutoCAD Color No.
17	0.70	0.028	Halftone	67	5	17
18	0.35	0.014	Halftone	75	2	18
19	2.00	0.079	Black	83	15	19
20	0.18	0.007	Rust	6	0	20
21	0.25	0.010	Rust	30	1	21
22	0.35	0.014	Rust	22	2	22
23	0.50	0.020	Rust	46	3	23
24	0.70	0.028	Rust	38	5	24
25	1.00	0.039	Rust	62	7	25
26	1.40	0.055	Rust	54	10	26
27	0.70	0.028	Rust	78	5	27
28	0.35	0.014	Rust	70	2	28
29	2.00	0.079	Rust	94	15	29
30	0.18	0.007	Black	86	0	30
31	0.25	0.010	Black	110	1	31
32	0.35	0.014	Black	102	2	32
33	0.50	0.020	Black	126	3	33
34	0.70	0.028	Black	118	5	34
35	1.00	0.039	Black	142	7	35
36	1.40	0.055	Black	134	10	36
37	0.70	0.028	Halftone	158	5	37
38	0.35	0.014	Halftone	150	2	38
39	2.00	0.079	Black	174	15	39
40	0.18	0.007	Gold	166	0	40
41	0.25	0.010	Gold	190	1	41
42	0.35	0.014	Gold	182	2	42
43	0.50	0.020	Gold	206	3	43
44	0.70	0.028	Gold	198	5	44
45	1.00	0.039	Gold	222	7	45
46	1.40	0.055	Gold	214	10	46
47	0.70	0.028	Gold	238	5	47
48	0.35	0.014	Gold	230	2	48

Color Number	Pen Plotter mm	Laser in.	Plot Color	MicroStation Color No.	Linewt.	AutoCAD Color No.
49	2.00	0.079	Gold	251	15	49
50	0.18	0.007	Black	20	0	50
51	0.25	0.010	Black	28	1	51
52	0.35	0.014	Black	36	2	52
53	0.50	0.020	Black	44	3	53
54	0.70	0.028	Black	52	5	54
55	1.00	0.039	Black	60	7	55
56	1.40	0.055	Black	68	10	56
57	0.70	0.028	Halftone	76	5	57
58	0.35	0.014	Halftone	84	2	58
59	2.00	0.079	Black	92	15	59
60	0.18	0.007	Olive	100	0	60
61	0.25	0.010	Olive	108	1	61
62	0.35	0.014	Olive	116	2	62
63	0.50	0.020	Olive	124	3	63
64	0.70	0.028	Olive	132	5	64
65	1.00	0.039	Olive	140	7	65
66	1.40	0.055	Olive	148	10	66
67	0.70	0.028	Olive	156	5	67
68	0.35	0.014	Olive	164	2	68
69	2.00	0.079	Olive	172	15	69
70	0.18	0.007	Black	180	0	70
71	0.25	0.010	Black	188	1	71
72	0.35	0.014	Black	196	2	72
73	0.50	0.020	Black	204	3	73
74	0.70	0.028	Black	212	5	74
75	1.00	0.039	Black	220	7	75
76	1.40	0.055	Black	228	10	76
77	0.70	0.028	Halftone	236	5	77
78	0.35	0.014	Halftone	244	2	78
79	2.00	0.079	Black	252	15	79
80	0.18	0.007	Green	11	0	80

Color Number	Pen Plotter mm	Laser in.	Plot Color	MicroStation Color No.	Linewt.	AutoCAD Color No.
81	0.25	0.010	Green	26	1	81
82	0.35	0.014	Green	18	2	82
83	0.50	0.020	Green	42	3	83
84	0.70	0.028	Green	34	5	84
85	1.00	0.039	Green	58	7	85
86	1.40	0.055	Green	50	10	86
87	0.70	0.028	Green	74	5	87
88	0.35	0.014	Green	66	2	88
89	2.00	0.079	Green	90	15	89
90	0.18	0.007	Black	82	0	90
91	0.25	0.010	Black	106	1	91
92	0.35	0.014	Black	98	2	92
93	0.50	0.020	Black	122	3	93
94	0.70	0.028	Black	114	5	94
95	1.00	0.039	Black	138	7	95
96	1.40	0.055	Black	130	10	96
97	0.70	0.028	Halftone	154	5	97
98	0.35	0.014	Halftone	146	2	98
99	2.00	0.079	Black	170	15	99
100	0.18	0.007	Forest Green	162	0	100
101	0.25	0.010	Forest Green	186	1	101
102	0.35	0.014	Forest Green	178	2	102
103	0.50	0.020	Forest Green	202	3	103
104	0.70	0.028	Forest Green	194	5	104
105	1.00	0.039	Forest Green	218	7	105
106	1.40	0.055	Forest Green	210	10	106
107	0.70	0.028	Forest Green	234	5	107
108	0.35	0.014	Forest Green	226	2	108
109	2.00	0.079	Forest Green	250	15	109
110	0.18	0.007	Black	242	0	110
111	0.25	0.010	Black	246	1	111
112	0.35	0.014	Black	247	2	112

Color Number	Pen Plotter mm	Laser in.	Plot Color	MicroStation Color No.	Linewt.	AutoCAD Color No.
113	0.50	0.020	Black	16	3	113
114	0.70	0.028	Black	32	5	114
115	1.00	0.039	Black	48	7	115
116	1.40	0.055	Black	64	10	116
117	0.70	0.028	Halftone	80	5	117
118	0.35	0.014	Halftone	96	2	118
119	2.00	0.079	Black	112	15	119
120	0.18	0.007	Teal	12	0	120
121	0.25	0.010	Teal	15	1	121
122	0.35	0.014	Teal	23	2	122
123	0.50	0.020	Teal	31	3	123
124	0.70	0.028	Teal	39	5	124
125	1.00	0.039	Teal	47	7	125
126	1.40	0.055	Teal	55	10	126
127	0.70	0.028	Teal	63	5	127
128	0.35	0.014	Teal	71	2	128
129	2.00	0.079	Teal	79	15	129
130	0.18	0.007	Black	87	0	130
131	0.25	0.010	Black	95	1	131
132	0.35	0.014	Black	103	2	132
133	0.50	0.020	Black	111	3	133
134	0.70	0.028	Black	119	5	134
135	1.00	0.039	Black	127	7	135
136	1.40	0.055	Black	135	10	136
137	0.70	0.028	Halftone	143	5	137
138	0.35	0.014	Halftone	151	2	138
139	2.00	0.079	Black	159	15	139
140	0.18	0.007	Cyan	167	0	140
141	0.25	0.010	Cyan	175	1	141
142	0.35	0.014	Cyan	183	2	142
143	0.50	0.020	Cyan	191	3	143
144	0.70	0.028	Cyan	199	5	144

Color Number	Pen Plotter mm	Laser in.	Plot Color	MicroStation Color No.	Linewt.	AutoCAD Color No.
145	1.00	0.039	Cyan	207	7	145
146	1.40	0.055	Cyan	215	10	146
147	0.70	0.028	Cyan	223	5	147
148	0.35	0.014	Cyan	231	2	148
149	2.00	0.079	Cyan	239	15	149
150	0.18	0.007	Black	40	0	150
151	0.25	0.010	Black	72	1	151
152	0.35	0.014	Black	88	2	152
153	0.50	0.020	Black	104	3	153
154	0.70	0.028	Black	136	5	154
155	1.00	0.039	Black	152	7	155
156	1.40	0.055	Black	184	10	156
157	0.70	0.028	Halftone	216	5	157
158	0.35	0.014	Halftone	232	2	158
159	2.00	0.079	Black	248	15	159
160	0.18	0.007	Blue	17	0	160
161	0.25	0.010	Blue	25	1	161
162	0.35	0.014	Blue	33	2	162
163	0.50	0.020	Blue	41	3	163
164	0.70	0.028	Blue	49	5	164
165	1.00	0.039	Blue	57	7	165
166	1.40	0.055	Blue	65	10	166
167	0.70	0.028	Blue	73	5	167
168	0.35	0.014	Blue	81	2	168
169	2.00	0.079	Blue	89	15	169
170	0.18	0.007	Black	97	0	170
171	0.25	0.010	Black	105	1	171
172	0.35	0.014	Black	113	2	172
173	0.50	0.020	Black	121	3	173
174	0.70	0.028	Black	129	5	174
175	1.00	0.039	Black	137	7	175
176	1.40	0.055	Black	145	10	176

Color Number	Pen Plotter mm	Laser in.	Plot Color	MicroStation Color No.	Linewt.	AutoCAD Color No.
177	0.70	0.028	Halftone	153	5	177
178	0.35	0.014	Halftone	161	2	178
179	2.00	0.079	Black	169	15	179
180	0.18	0.007	Navy	177	0	180
181	0.25	0.010	Navy	185	1	181
182	0.35	0.014	Navy	193	2	182
183	0.50	0.020	Navy	201	3	183
184	0.70	0.028	Navy	209	5	184
185	1.00	0.039	Navy	217	7	185
186	1.40	0.055	Navy	225	10	186
187	0.70	0.028	Navy	233	5	187
188	0.35	0.014	Navy	241	2	188
189	2.00	0.079	Navy	249	15	189
190	0.18	0.007	Black	245	0	190
191	0.25	0.010	Black	128	1	191
192	0.35	0.014	Black	144	2	192
193	0.50	0.020	Black	160	3	193
194	0.70	0.028	Black	176	5	194
195	1.00	0.039	Black	192	7	195
196	1.40	0.055	Black	208	10	196
197	0.70	0.028	Halftone	224	5	197
198	0.35	0.014	Halftone	240	2	198
199	2.00	0.079	Black	254	15	199
200	0.18	0.007	Purple	13	0	200
201	0.25	0.010	Purple	29	1	201
202	0.35	0.014	Purple	21	2	202
203	0.50	0.020	Purple	45	3	203
204	0.70	0.028	Purple	37	5	204
205	1.00	0.039	Purple	61	7	205
206	1.40	0.055	Purple	53	10	206
207	0.70	0.028	Purple	77	5	207
208	0.35	0.014	Purple	69	2	208

Color Number	Pen Plotter mm	Laser in.	Plot Color	MicroStation Color No.	Linewt.	AutoCAD Color No.
209	2.00	0.079	Purple	93	15	209
210	0.18	0.007	Black	85	0	210
211	0.25	0.010	Black	109	1	211
212	0.35	0.014	Black	101	2	212
213	0.50	0.020	Black	125	3	213
214	0.70	0.028	Black	117	5	214
215	1.00	0.039	Black	141	7	215
216	1.40	0.055	Black	133	10	216
217	0.70	0.028	Halftone	157	5	217
218	0.35	0.014	Halftone	149	2	218
219	2.00	0.079	Black	173	15	219
220	0.18	0.007	Magenta	165	0	220
221	0.25	0.010	Magenta	189	1	221
222	0.35	0.014	Magenta	181	2	222
223	0.50	0.020	Magenta	205	3	223
224	0.70	0.028	Magenta	197	5	224
225	1.00	0.039	Magenta	221	7	225
226	1.40	0.055	Magenta	213	10	226
227	0.70	0.028	Magenta	237	5	227
228	0.35	0.014	Magenta	229	2	228
229	2.00	0.079	Magenta	253	15	229
230	0.18	0.007	Black	91	0	230
231	0.25	0.010	Black	99	1	231
232	0.35	0.014	Black	107	2	232
233	0.50	0.020	Black	115	3	233
234	0.70	0.028	Black	123	5	234
235	1.00	0.039	Black	131	7	235
236	1.40	0.055	Black	139	10	236
237	0.70	0.028	Halftone	147	5	237
238	0.35	0.014	Halftone	155	2	238
239	2.00	0.079	Black	163	15	239
240	0.18	0.007	Red	171	0	240

Color Number	Pen Plotter mm	Laser in.	Plot Color	MicroStation Color No.	Linewt.	AutoCAD Color No.
241	0.25	0.010	Red	179	1	241
242	0.35	0.014	Red	187	2	242
243	0.50	0.020	Red	195	3	243
244	0.70	0.028	Red	203	5	244
245	1.00	0.039	Red	211	7	245
246	1.40	0.055	Red	219	10	246
247	0.70	0.028	Red	227	5	247
248	0.35	0.014	Red	235	2	248
249	2.00	0.079	Red	243	15	249
250	0.25	0.010	Halftone	8	1	250
251	0.35	0.014	Halftone	200	2	251
252	0.50	0.020	Halftone	168	3	252
253	0.70	0.028	Halftone	120	5	253
254	1.00	0.039	Halftone	56	7	254
255	2.00	0.079	Halftone	24	15	255

The CSI Layer Standards

C

The Construction Specifications Institute originally created MasterFormat as a logical way of categorizing construction material; it was never meant for layers. Some CAD users, however, found the 17 divisions useful as a layering system.

MASTERFORMAT

Division 0 : Introductory Information, Bidding and Contracting Requirements

00001	Project Title Page
00005	Certifications Page
00007	Seals Page
00010	Table of Contents
00015	List of Drawings
00020	List of Schedules
00100	Bid Solicitation
00200	Instructions to Bidders
00300	Information Available to Bidders
00400	Bid Forms and Supplements
00490	Bidding Addenda
00500	Agreement
00600	Bonds and Certificates
00700	General Conditions
00800	Supplementary Conditions
00900	Modifications

APPENDIX

The Construction Specifications Institute has created two systems for organizing information for construction projects:

- **MasterFormat** is a layer system based upon 17 divisions.
- **UniFormat** is a mixed letter and number system based upon seven divisions.
- **US Coast Guard** is an adaptation of the MasterFormat, with two added divisions.

These systems are suitable for use as a layer naming system. This appendix reprints the "light" versions of the CSI systems, as well as the full United States Coast Guard layer specification.

MasterFormat (cont'd)

Division 1: General Requirements

01100	Summary
01200	Price and Payment Procedures
01300	Administrative Requirements
01400	Quality Requirements
01500	Temporary Facilities and Controls
01600	Product Requirements
01700	Execution Requirements
01800	Facility Operation
01900	Facility Decommissioning

Division 2: Site Construction

02050	Basic Site Materials and Methods
02100	Site Remediation
02200	Site Preparation
02300	Earthwork
02400	Tunneling, Boring, and Jacking
02450	Foundation and Load-bearing Elements
02500	Utility Services
02600	Drainage and Containment
02700	Bases, Ballasts, Pavements, and Appurtenances
02800	Site Improvements and Amenities
02900	Planting
02950	Site Restoration and Rehabilitation

Division 3: Concrete

03050	Basic Concrete Materials and Methods
03100	Concrete Forms and Accessories
03200	Concrete Reinforcement
03300	Cast-in-Place Concrete
03400	Precast Concrete
03500	Cementitious Decks and Underlayment
03600	Grouts
03700	Mass Concrete
03900	Concrete Restoration and Cleaning

Division 4: Masonry

04050	Basic Masonry Materials and Methods
04200	Masonry Units
04400	Stone
04500	Refractories
04600	Corrosion-Resistant Masonry
04700	Simulated Masonry
04800	Masonry Assemblies
04900	Masonry Restoration and Cleaning

Division 5: Metals

05050	Basic Metal Materials and Methods
05100	Structural Metal Framing
05200	Metal Joists
05300	Metal Deck
05400	Cold-Formed Metal Framing
05500	Metal Fabrications
05600	Hydraulic Fabrications
05650	Railroad Track and Accessories
05700	Ornamental Metal
05800	Expansion Control
05900	Metal Restoration and Cleaning

Division 6: Wood and Plastics

06050	Basic Wood and Plastic Materials and Methods
06100	Rough Carpentry
06200	Finish Carpentry
06400	Architectural Woodwork
06500	Structural Plastics
06600	Plastic Fabrications
06900	Wood and Plastic Restoration and Cleaning

Division 7: Thermal and Moisture Protection

07050	Basic Thermal and Moisture Protection Materials and Methods
07100	Dampproofing and Waterproofing
07200	Thermal Protection
07300	Shingles, Roof Tiles, and Roof Coverings
07400	Roofing and Siding Panels
07500	Membrane Roofing
07600	Flashing and Sheet Metal
07700	Roof Specialties and Accessories
07800	Fire and Smoke Protection
07900	Joint Sealers

RESOURCES

To obtain the full MasterFormat or UniFormat systems, contact CSI at:

The Construction Specifications Institute
99 Canal Center Plaza
Suite 300
Alexandria VA 22314
Tel. (800) 689-2900

Construction Specifications Canada
100 Lombard Street
Suite 200
Toronto, Ontario
M5C 1M3
Tel. (416) 777-2198

Email
membcustsrv@csinet.org
Web site:
www.csinet.org

For more information about the Coast Guard's adaptation, contact:
US Coast Guard
Civil Engineering
Technology Center
At CEU Cleveland, Ohio
Tel. (216) 522-3934 x347
www.uscg.mil/mlclant/cetc/cadprods.htm

The CSI Layer Standards

MasterFormat (cont'd)

Division 8: Doors and Windows

08050	Basic Door and Window Materials and Methods
08100	Metal Doors and Frames
08200	Wood and Plastic Doors
08300	Specialty Doors
08400	Entrances and Storefronts
08500	Windows
08600	Skylights
08700	Hardware
08800	Glazing
08900	Glazed Curtain Wall

Division 9: Finishes

09050	Basic Finish Materials and Methods
09100	Metal Support Assemblies
09200	Plaster and Gypsum Board
09300	Tile
09400	Terrazzo
09500	Ceilings
09600	Flooring
09700	Wall Finishes
09800	Acoustical Treatment
09900	Paints and Coatings

Division 10: Specialties

10100	Visual Display Boards
10150	Compartments and Cubicles
10200	Louvers and Vents
10240	Grilles and Screens
10250	Service Walls
10260	Wall and Corner Guards
10270	Access Flooring
10290	Pest Control
10300	Fireplaces and Stoves
10340	Manufactured Exterior Specialties
10350	Flagpoles
10400	Identification Devices
10450	Pedestrian Control Devices
10500	Lockers
10520	Fire Protection Specialties
10530	Protective Covers

10550	Postal Specialties
10600	Partitions
10670	Storage Shelving
10700	Exterior Protection
10750	Telephone Specialties
10800	Toilet, Bath, and Laundry Accessories
10880	Scales
10900	Wardrobe and Closet Specialties

Division 11: Equipment

11010	Maintenance Equipment
11020	Security and Vault Equipment
11030	Teller and Service Equipment
11040	Ecclesiastical Equipment
11050	Library Equipment
11060	Theater and Stage Equipment
11070	Instrumental Equipment
11080	Registration Equipment
11090	Checkroom Equipment
11100	Mercantile Equipment
11110	Commercial Laundry and Dry Cleaning Equipment
11120	Vending Equipment
11130	Audio-Visual Equipment
11140	Vehicle Service Equipment
11150	Parking Control Equipment
11160	Loading Dock Equipment
11170	Solid Waste Handling Equipment
11190	Detention Equipment
11200	Water Supply and Treatment Equipment
11280	Hydraulic Gates and Valves
11300	Fluid Waste Treatment and Disposal Equipment
11400	Food Service Equipment
11450	Residential Equipment
11460	Unit Kitchens
11470	Darkroom Equipment
11480	Athletic, Recreational, and Therapeutic Equipment
11500	Industrial and Process Equipment
11600	Laboratory Equipment
11650	Planetarium Equipment
11660	Observatory Equipment
11680	Office Equipment
11700	Medical Equipment

MasterFormat (cont'd)

11780	Mortuary Equipment
11850	Navigation Equipment
11870	Agricultural Equipment
11900	Exhibit Equipment

Division 12: Furnishings

12050	Fabrics
12100	Art
12300	Manufactured Casework
12400	Furnishings and Accessories
12500	Furniture
12600	Multiple Seating
12700	Systems Furniture
12800	Interior Plants and Planters
12900	Furnishings Repair and Restoration

Division 13: Special Construction

13010	Air-Supported Structures
13020	Building Modules
13030	Special Purpose Rooms
13080	Sound, Vibration, and Seismic Control
13090	Radiation Protection
13100	Lightning Protection
13110	Cathodic Protection
13120	Pre-Engineered Structures
13150	Swimming Pools
13160	Aquariums
13165	Aquatic Park Facilities
13170	Tubs and Pools
13175	Ice Rinks
13185	Kennels and Animal Shelters
13190	Site-Constructed Incinerators
13200	Storage Tanks
13220	Filter Underdrains and Media
13230	Digester Covers and Appurtenances
13240	Oxygenation Systems
13260	Sludge Conditioning Systems
13280	Hazardous Material Remediation
13400	Measurement and Control Instrumentation
13500	Recording Instrumentation
13550	Transportation Control Instrumentation
13600	Solar and Wind Energy Equipment

13700 Security Access and Surveillance
13800 Building Automation and Control
13850 Detection and Alarm
13900 Fire Suppression

Division 14: Conveying Systems

14100 Dumbwaiters
14200 Elevators
14300 Escalators and Moving Walks
14400 Lifts
14500 Material Handling
14600 Hoists and Cranes
14700 Turntables
14800 Scaffolding
14900 Transportation

Division 15: Mechanical

15050 Basic Mechanical Materials and Methods
15100 Building Services Piping
15200 Process Piping
15300 Fire Protection Piping
15400 Plumbing Fixtures and Equipment
15500 Heat-Generation Equipment
15600 Refrigeration Equipment
15700 Heating, Ventilating, and Air Conditioning Equipment
15800 Air Distribution
15900 HVAC Instrumentation and Controls
15950 Testing, Adjusting, and Balancing

Division 16: Electrical

16050 Basic Electrical Materials and Methods
16100 Wiring Methods
16200 Electrical Power
16300 Transmission and Distribution
16400 Low-Voltage Distribution
16500 Lighting
16700 Communications
16800 Sound and Video

UNIFORMAT

10: Project Description

1010	Project Summary
1020	Project Program
1030	Existing Conditions
1040	Owner's Work
1050	Funding

20: Proposal, Bidding, and Contracting

2010	Delivery Method
2020	Qualifications Requirements
2030	Proposal Requirements
2040	Bid Requirements
2050	Contracting Requirements

30: Cost Summary

3010	Elemental Cost Estimate
3020	Assumptions and Qualifications
3030	Allowances
3040	Alternates
3050	Unit Prices

A: Substructure

A10 Foundations
A1010	Standard Foundations
A1020	Special Other Foundations
A1030	Slabs on Grade

A20 Basement Construction
A2010	Basement Excavation
A2020	Basement Walls

B: Shell

B10 Superstructure
B1010	Floor Construction
B1020	Roof Construction

B20	**Exterior Enclosure**
B2010	Exterior Walls
B2020	Exterior Windows
B2030	Exterior Doors
B30	**Roofing**
B3010	Roof Coverings
B3020	Roof Openings

C: Interiors

C10	**Interior Construction**
C1010	Partitions
C1020	Interior Doors
C1030	Fittings Specialties
C20	**Stairs**
C2010	Stair Construction
C2020	Stair Finishes
C30	**Interior Finishes**
C3010	Wall Finishes
C3020	Floor Finishes
C3030	Ceiling Finishes

D: Services

D10	**Conveying Systems**
D1010	Elevators and Lifts
D1020	Escalators and Moving Walks
D1030	Materials Handling
D1090	Other Conveying Systems
D20	**Plumbing**
D2010	Plumbing Fixtures
D2020	Domestic Water Distribution
D2030	Sanitary Waste
D2040	Rain Water Drainage
D2090	Other Plumbing Systems

UniFormat (cont'd)

D30	**Heating, Ventilating, Air Conditioning (HVAC)**
D3010	Fuel Energy Supply Systems
D3020	Heat Generation Systems
D3030	Heat Rejection Systems Refrigeration
D3040	Heat HVAC Distribution Systems
D3050	Heat Transfer Terminal and Packaged Units
D3060	HVAC Instrumentation and Controls
D3070	HVAC Systems Testing, Adjusting, and Balancing
D3090	Other Special HVAC Systems and Equipment
D40	**Fire Protection Systems**
D4010	Fire Protection Sprinklers Systems
D4020	Standpipes and Hose Systems
D4030	Fire Protection Specialties
D4090	Other Fire Protection Systems
D50	**Electrical Systems**
D5010	Electrical Service and Distribution
D5020	Lighting and Branch Wiring
D5030	Communications and Security Systems
D5040	Special Electrical Systems
D5050	Electrical Controls and Instrumentation
D5060	Electrical Testing
D5090	Other Electrical Systems

E: Equipment and Furnishings

E10	**Equipment**
E1010	Commercial Equipment
E1020	Institutional Equipment
E1030	Vehicular Equipment
E1090	Other Equipment
E20	**Furnishings**
E2010	Fixed Furnishings
E2020	Movable Furnishings

F: Special Construction and Demolition

F10	**Special Construction**
F1010	Special Structures
F1020	Integrated Construction
F1030	Special Construction Systems
F1040	Special Facilities
F1050	Special Controls and Instrumentation

F20   **Selective Demolition**
F2010   Building Elements Demolition
F2020   Hazardous Components Abatement

G: Building Sitework

G10   **Site Preparation**
G1010   Site Clearing
G1020   Site Demolition and Relocations
G1030   Site Earthwork
G1040   Hazardous Waste Remediation

G20   **Site Improvements**
G2010   Roadways
G2020   Parking Lots
G2030   Pedestrian Paving
G2040   Site Development
G2050   Landscaping

G30   **Site Civil/Mechanical Utilities**
G3010   Water Supply
G3020   Sanitary Sewer
G3030   Storm Sewer
G3040   Heating Distribution
G3050   Cooling Distribution
G3060   Fuel Distribution
G3090   Other Site Mechanical Utilities

G40   **Site Electrical Utilities**
G4010   Electrical Distribution
G4020   Site Lighting
G4030   Site Communications and Security
G4090   Other Site Electrical Utilities

G90   **Other Site Construction**
G9010   Service Tunnels
G9090   Other Site Systems

UniFormat (cont'd)

Z: General

Z10	**General Requirements**
Z1010	Administration
Z1020	Procedural, General Requirements, and Quality Requirements
Z1030	Temporary Facilities and Temporary Controls
Z1040	Project Closeout
Z1050	Permits, Insurance, and Bonds
Z1060	Fee
Z20	**Bidding Requirements, Contract Forms, and Conditions Contingencies**
Z2010	Bidding Requirements Design Contingency
Z2020	Contract Forms Escalation Contingency
Z2030	Conditions Construction Contingency
Z90	**Project Cost Estimate**
Z9010	Lump Sum
Z9020	Unit Prices
Z9030	Alternates/Alternatives

US COAST GUARD MASTER LISTING

Division 01: Field Engineering

01050	Surveyor Grid
01051	Property Line
01052	Tract
01053	Easement
01190	Environmental

Division 02: Sitework

02200	Soil
02276	Retaining Wall
02280	Soil Boring
02350	Piling, Caisson
02450	Railroad
02480	Water Craft
02515	Asphalt, Pavement
02516	Parking Lot
02517	Roadway
02645	Fire Hydrant
02670	Well
02720	Storm Drainage
02770	Ponds, Reservoirs
02786	Pole, Tower
02830	Fence, Gate
02880	Monument
02950	Landscaping
02980	Landscaping Accessories

Division 03: Concrete

03010	Concrete
03011	Gravel
03012	Sand
03100	Sidewalk
03200	Wire Mesh, Re-Bar
03400	Precast Concrete
03600	Grout
03650	Foundation

The CSI Layer Standards

USCG Master Listing
(cont'd)

Division 04: Masonry

04200	Brick Masonry Unit
04220	Concrete Masonry Unit
04400	Stone

Division 05: Metals

05010	Aluminum
05015	Steel
05016	Iron
05020	Brass
05021	Copper
05022	Nickel
05030	Coating
05050	Fastener
05100	Main Column
05102	Main Beam
05160	Truss
05200	Miscellaneous Structural
05202	Bracing
05515	Ladder
05532	Grating
05535	Metal Decking
05712	Stair
05720	Handrail

Division 06: Wood and Plastics

06010	Wood
06190	Pre-Fab Joist, Truss
06220	Millwork
06430	Wood Stairs, Handrails
06432	Wood Ladders
06500	Plastic
06610	Fiberglass

Division 07: Thermal and Moisture Protection

07100	Waterproofing
07200	Insulation
07420	Exterior Wall
07460	Siding
07500	Roofing
07650	Flashing

Division 08: Doors and Windows

08100	Door
08360	Overhead Door
08500	Window
08730	Weather-Stripping
08810	Glass

Division 09: Finishes

09110	Interior Wall
09130	Ceiling
09250	Plaster
09300	Tile
09500	Acoustical
09650	Rubber
09700	Flooring
09900	Building Shading

Division 10: Specialties

10350	Flagpoles
10630	Portable Partitions
10880	Lockers

Division 11: Equipment

11120	Vending Equipment
11140	Vehicle Service Equipment
11160	Loading Dock Equipment
11400	Food Service Equipment
11452	Appliances
11528	Paint Spray Equipment
11530	Safety Equipment

Division 12: Furnishings

12050	Cloth
12302	Casework
12600	Furniture
12605	Screen

Division 13: Special Construction

13120	Pre-Engineered Structure
13145	Mezzanine
13152	Swimming Pools, Equipment
13200	Liquid, Gas Storage

The CSI Layer Standards

USCG Master Listing (cont'd)

Division 14: Conveying Systems

14200	Elevators
14205	Vehicle
14520	Emergency Vehicle
14550	Conveyers
14600	Hoists, Cranes

Division 15: Mechanical

15070	Water: Utility
15071	Water: Process
15072	Water: Fire Protection
15073	Liquid: Process
15165	Compressor
15202	Filtration Equipment
15300	Fire Protection System
15320	Sprinkler System
15400	Plumbing
15435	Drain: Floor, Roof
15436	Drain: Process
15440	Plumbing Fixtures
15444	Safety Shower
15455	Drinking Fountain Water Cooler
15470	Unit Storage Tanks
15481	Compressed Air
15488	Gas: Process
15489	Gas: Hazardous
15490	Gas: Inert
15491	Fuel: Oil
15492	Fuel: Gas
15493	Fuel: Jet
15494	Fuel: Liquid
15500	HVAC
15525	Steam Condensate
15555	Boilers, Water Heaters
15620	Fuel Fire Heaters
15780	Packaged Air Conditioners
15835	Convectors, Radiators
15860	Air Handling Fans
15877	Fume Exhaust Equipment
15885	Air Cleaning Equipment
15890	Duct Work
15942	Registers, Grilles, Diffusers
15950	Mech. Controls, Instrumentation

Division 16: Electrical

16100	Electrical
16110	Raceways, Conduits
16120	Wires, Cables
16140	Wiring Devices
16180	Motors
16200	Power Generator
16390	Grounding Devices
16420	Service Opening
16500	Lighting
16530	Aviation Lighting
16532	Navigation Lighting
16700	Communication
16710	Computer
16720	Alarm Detection System
16740	Telecommunication
16880	Electric Radiant Heaters

Division 20: Reference

20100	Centerline (Center)
20102	Centerline (Center3)
20105	Col. Balloon, Col. Centerline (Center2)
20200	Phantom Line (Phantom)
20202	Phantom Line (Phantom3)
20205	Match Line (Phantom2)
20300	Broken Line (Hidden)
20302	Broken Line (Hidden3)
20305	Broken Line (Dashed)
20400	Solid Line, Continuous
20500	Leaders, Leader Text
20600	Dimensions, Notes, Text
20700	Miscellaneous
20800	Revision Cloud, Tag
20900	Crosshatch

Index

A

A through E 108
A/E/C CADD symbol libraries 45
A0 through A4 107-100
after templates are created 127
aging of software 192
AIA, 55
    color guidelines 77
    discipline codes 29
    file naming convention 29
    layer groups and subgroups 57
    layer major categories 56
    layer modifiers 57
    standards 81
American Institute of Architects: *see* AIA
anatomy of a dimension 101
ANSI sheet sizes 108
archiving,
    drawings 141
    travails of 20
are vendors helping CAD managers? 24
ASP-based systems 189
assigning colors 69
AutoCAD,
    colors 74
    printing layers 51
automating repetitive tasks 9

B

beginnings of CAD 196
blocks: *see* symbols
borders 111
budget,
    CAD department 4
    capital expenses 6
    justifying new systems 7
    representative software costs 6
    representative software costs 6
    service pricing 165
    services 159

C

CAD,
    beginnings of 196
    department budget 4
    for Linux 249
    for Macintosh 248
    for Windows 243
    future of 190
    processes 2
    recent history of 195
    standards manual, writing your 129
    systems 1
    underutilization 13
    vendors 243
    without layers 48
CAD manager,
    are vendors helping? 24
    issues 15
    justifying new systems 7
    keeping up-to-date 3
    ongoing expenses 6
    representative software costs 6
    resources for 241
    role of 1
    staffing levels 4
    what's a CAD manager worth? 4
California Department of Transportation: *see* CalTrans

CalTrans,
    drawing data levels   63
    file naming convention   34
    text standard   83
capital expenses   6
case study: all-digital drawing   154
case,
    against CAD management   17
    for color   69
    for monochrome   70
cells: *see* symbols
colors,   69
    computer numbering systems   73
    depth   145
    for AutoCAD users   74
    how CAD works with   73
    matching to pens   74
    pen table   253
    standards   77
    where to assign   74
common standards organizations   50
compression issues   145
computer-aided design: *see* CAD
computer color numbering systems   73
conflict of the disciplines   53
Construction Specifications Institute: *see* CSI
converting drawings   149
coordination   8
creating,
    symbols   27
    template drawing   116
CSI,   81
    drawing sheet specification   112
    layer standards   263
    MasterFormat layers   58
    nomenclature   116
    placing sheets in order   113
    single design model   113
    standards for sheet identification   114
    uniform drawing system   31
    UniFormat layers   60

D

design file: *see* drawing
determining scale factor   98
development of the CAD font   82
digitizing and scanning 11
dimensions,   97
    3D   100
    anatomy of   101
    standards   102
    with CAD   100
DIN (German) sheet sizes   110
disadvantages of CAD   18
discipline codes   29
division of work   7
do CAD brands matter anymore?   15
dots per inch   144
downside to objects   186
drawings,
    archiving   141
    converting   149
    DWG specification   171
    eliminating the paper trail   154
    future of   182
    originals not accurate   150
    sizes   106
    the all-digital   154
    working with paper   139
DWG,
    file specification   171
    format and its future   169

E

effect of resolution on scan quality   146
eliminating the paper trail   154
extranets   157

F

factor, determining the scale   98
file extensions   36
file names,
    AIA convention   29
    CalTrans convention   34
    industry standards for   28
    simple convention   27
file,
    DWG specification   171
    future of formats   182
font,   79
    AIA standard   81
    CalTrans standard   83
    CSI standard   81
    development of the CAD font   82
    imperial text heights   85
    metric text heights   85
    TSTC text standard   86
    USACE text standard   85
    USCG text standard   84
rom files to central databases   184
future,
    of CAD   190
    of file formats   182

G

gallery of Web-based services   161
German (DIN) sheet sizes   110
GIF   148
Graphic Interchange Format: *see* GIF

H

hardware,
    procurement   10
    selection   9
    whim of fashion   8
hardwired versus customized linetypes   87

hatch patterns 94
hatching: *see also* patterns
history of CAD, 195
    year 1995 197
    year 1996 202
    year 1997 208
    year 1998 216
    year 1999 224
    year 2000 230
    year 2001 236
how CAD works with colors 73
how to,
    create a symbol library 38
    name a layer 49
human quirks 54

I

imperial,
    ANSI sheet sizes 108
    text heights 85
industry standards for file names 28
Internet 10
is CAD dead? 19
ISO,
    metric sheet sizes 107
    layer standard 64
    layers, mandatory part 64
    layers, optional part 66
International Organization of Standards: *see* ISO

J

JIS (Japanese) sheet sizes 109
Joint Photographic Experts Group: *see* JPEG
JPEG,
    2000 and JPIG2 148
    why you shouldn't use 148
justifying new systems 7

L

layer names,
    conventions   47
    for in-house drawings   52
layers,
    AIA groups and subgroups   57
    AIA major categories   56
    AIA modifiers   57
    American Institute of Architects   55
    CAD package without   48
    CalTrans   63
    conflict of the disciplines   53
    CSI MasterFormat layers   58
    CSI standards   263
    CSI UniFormat layers   60
    how to name   49
    human quirks   54
    in CAD   47
    ISO   64
    limitations of CAD systems   53
    printing AutoCAD   51
    struggle to create standards   53
    US Coast Guard   61
    what are?   47
levels: *see* layers
limitations of CAD systems   53
line widths   91
linetypes,   79, 87
    hardwired versus. customized 87
    one-dimensional versus two-dimensional 87
    scaling 88
    software versus hardware 88
    standards   89
    TSTC standard   90
    USCG standard   89
lineweights and colors   92
lineweights: *see also* linewidths

linewidth, 79
    MicroStation 91
    National CAD Standards line widths 92
    standards 91
    TSTC line width standard 92
    USACE line weight standard 93
linewidth: *see also* line weight
Linux CAD 249

M

Macintosh CAD 248
manual, writing your CAD standards 129
MasterFormat, 263
    CSI layers 58
matching colors to pens 74
metric,
    ISO sheet sizes 107
    text heights 85
MicroStation pen control 75
model versus sheet 65
monochrome, case for 70

N

naming drawings 27
National CAD Standards line widths 92
nomenclature 116

O

objects,
    downside to 186
    in CAD 182
    working with 185
one-dimensional versus two-dimensional linetypes 87
operating system 11
original drawings are not accurate 150
other UDS specs 31

outsourcing,   157
    project management   160
overcoming politics and whims   9

P

paper drawings, working with   139
paper space: *see* sheet
partial conversion strategy   153
patterns,   79
    hatch   94
patterns: *see also* hatching
pens,
    color table   253
      matching colors to   74
      MicroStation control   75
placing sheets in order   113
plotting,   12
    yesteryear   76
politics of software releases   7
preparing the template drawing   117
pricing, service bureau   165
printer, scanner   144
printing AutoCAD layers   51
project life-cycle   12

R

raster issues and calculations   144
recent history of CAD   195
representative software costs   6
resources for CAD managers   241
role of the CAD manager   1

S

scale factors,   97
    determining the   98
    linetypes   88

scanners, 141
    color depth 145
    compression issues 145
    dots per inch 144
    effect of resolution on scan quality 146
    GIF 148
    JBIG2 and JPEG2000 148
    partial conversion strategy 153
    printer 144
    software 142
    storage 143
    why you shouldn't use JPEG 148
scanning and digitizing 11
seed file: *see* template
server-based systems 187
service bureau,
    pricing 165
    what is? 158
sheet sizes,
    ANSI 108
    DIN (German) 110
    ISO (metric) 107
    JIS (Japanese) 109
sheet versus model 65
simple file name convention 27
single design model 113
software, 10
    aging of 192
    scanner 142
    versus hardware linetypes 88
sources of symbols 44
specification, DWG file 171
stages,
    1. Convincing yourself 7
    2. Meeting upper management's objections 10
    3. Convincing your staff 12

standards,
- bodies   251
- color   77
- CSI drawing sheet specification   112
- CSI layer   263
- dimensioning   102
- drawings and templates   105
- for sheet identification   114
- hatch pattern   94
- linewidth   91
- manual, writing your   129
- organizations, most common   50
- text   80

starting points   241
stay or switch?   23
storage, scanner   143
strategies, layer
- 0. Do nothing   50
- 1. The simple plan   50
- 2. The plotter plan   50
- 3. The four-step plan   51
- 4. Do what your client says   52
- 5. Copy what works, make minor modifications   52

struggle to create layer standards   53
symbols,
- A/E/C CADD libraries   45
- creation summary   38
- how to create a library   38
- in CAD   38
- sources of   44
- what are?   37

system administration   8

T

template drawing,   105
- after created   127
- creating a   116
- creation steps   118
- preparing the   117
- what is a?   116

text,
- AIA and CSI standards   81
- CalTrans standard   83
- development of the CAD font   82
- imperial heights   85
- metric heights   85
- standards   80
- TSTC standard   86
- USACE standard   85
- USCG standard   84

tips,
- azimuth   118
- color numbering systems — what they mean   72
- difference between vector and raster   150
- fonts versus styles   81
- getting help with windows   6
- graphic scale   123
- helpful windows utilities   5
- layers versus levels   50
- light gray screening   70
- line weights and colors   92
- MicroStation line weights   91
- more than 24 bits   145
- optical versus software resolution   141
- overview of drawing settings   117
- phraseology   39
- place comments on a nonplotting layer   85
- speeding up text   82
- threshholding grayscale scans   140
- wildcarding layer names   55

title blocks   111
training   7
translation   8
travails of archiving   20
TSTC,
- line width standard   92
- linetype standard   90
- text standard   86

two editing solutions   152

U

UniFormat,
    CSI layers   60
    standard   270
US Coast Guard,
    color guidelines   77
    layers   61
    linetype standard   89
    Master Listing   275
    text standard   84
USACE,
    line weight standard   93
    text standard   85

V

vendors,
    CAD   243
    helping CAD managers?   24
visualization   8

W

whim of hardware fashion   8
Windows CAD   243
work area and ergonomics   7
working,
    with objects   185
    with paper drawings   139
writing your CAD standards manual   129

Y

Y38 problem   228

Colophon

The author at an early age with mother Ricarda Grabowski.

An earlier edition of this book was written in 1993 under the name *The Successful CAD Manager's Handbook* (Delmar Publishers). At the time, DOS and Unix were the primary operating systems for CAD software because Windows was considered too slow. Several of the CAD packages discussed in the first edition are no longer available, and pen plotters are no longer the most popular way to plot a CAD drawing.

The earlier edition was typeset using WordPerfect 5.1 for DOS, and output on a Hewlett-Packard LaserJet III outfitted with a LaserMaster board that juiced up the resolution from 300 dpi to 800 dpi. The 100 KB worth of files were sent to the publisher on a diskette by courier.

This edition of the *CAD Manager's Handbook* was typeset on PageMaker 6.5, with the output going direct to the publisher's 2400 dpi typesetter. The 275 MB worth of files were sent to the publisher by high-speed email, FTP, and burned on a CD-R and delivered by courier.

The body text is set in New Baskerville; heads and subheads in Tahoma; tables and sidebar text in Swiss Condensed; and captions were set in Times New Roman italic.

Objects and photographs were scanned in with the Epson Perfection 1200S SCSI scanner, usually at 200 dpi. Page proofs were printed on a networked Lexmark Optra R+ at 600 dpi. Iomega Zip and Jaz disks were used for backing up files every day.

Credits

Karl Davies provided the map of Sawmill Hill made with Visio.

Herbert Grabowski provided the historic photographs and his hand-drafted drawings.

Volker Mueller, Martyn Day, and Shane Beaman voiced their opionion of CAD management in the first chapter.

Evan Yares of the OpenDWG Alliance gave permission to reproduce an edited version of their DWG specification in chapter 11, along with the full specifcation on the CD-ROM.

Rob Berry of IMSI gave permission to use the drawings by TurboCAD.

Martyn Day of *CADdesk AEC* magazine allowed the reprint of his interview and article on the future of CAD file formats in chapter 11.

Adena Schutzberg of Tenlinks.Com permitted the reprint of her GIS news item in chapter 12.

"The numbers and titles used in this presentation are from *MasterFormat* (1995 edition) and from *UniFormat* (1998 edition) and are published by the Construction Specifications Institute (CSI) and Construction Specifications Canada (CSC), and are used with permission from CSI, 2001.

"For those interested in a more in-depth explanation of *MasterFormat* and *UniFormat* and their use in the construction industry contact:

The Construction Specifications Institute (CSI)
99 Canal Center Plaza, Suite 300
800-689-2900; 703-684-0300
CSINet URL: http://www.csinet.org

Notes